Scientific Lives

Scientific Lives

John Aubrey

Hesperus Classics

Hesperus Classics
Published by Hesperus Press Limited
19 Bulstrode Street, London W1U 2JN
www.hesperuspress.com

This collection is based on *Brief Lives*, edited by Andrew Clark, first published in 1898
This selection first published by Hesperus Press Limited, 2011

Designed and typeset by Fraser Muggeridge studio
Printed in Jordan by Jordan National Press

ISBN: 978-1-84391-169-2

CONTENTS

FOREWORD

On one of the peaceful empty days between Christmas and New Year in 1648, John Aubrey went hunting with some friends in Wiltshire. He had been summoned home from undergraduate studies in Oxford to the county of his birth and upbringing because his father had fallen seriously ill that Christmas Eve. Aubrey adored Wiltshire. As a child he had wandered often on Salisbury Plain and been fascinated by the mysterious megaliths at Stonehenge, but he did not know the countryside further north around Marlborough. It was from here that the hounds set off. The chase was long and led eventually through the village of Avebury, where 'the sight of those vast stones, of which I had never heard before', burst upon Aubrey. He reined back his horse and left the hunt to look more closely at the bank and ditch around the strange circles of enormous stones. His mind, from childhood imaginatively involved with the past, immediately began picturing how the great stones had looked 'in the old time'. He thought the circles must originally have been complete and erected by Druids. Lost to the world, Aubrey suddenly heard the hounds again and hastened off to overtake them at Kennett. But he went back many times to Avebury, once, years later, accompanied by Charles II, who ordered him to dig there for human remains. 'But I did not doe it', Aubrey remarked, without apology or explanation. Instead he made meticulous drawings of the historic remains. He was horrified to find they were still being pillaged and the stones broken up for new building material. He campaigned to stop the desecration, and what the world travels to Avebury to see today, is owed largely to Aubrey's antiquarian instincts.

Aubrey was elected an early fellow of the Royal Society in 1663, where, along with Samuel Pepys and John Evelyn, he discussed scientific innovation with eminent men of the age. Aubrey was more modest than most. He offered his theory that Avebury and Stonehenge were Druid temples as a 'probability'. 'This Inquiry, I must confess, is gropeing in the Dark: but although I have not brought it into a cleer light yet I can affirm that I have brought it from utter darkness, to a thin Mist, and have gone further in this Essay than anyone before me.' He was as right as he was wrong: we know now that the stone temples pre-dated the Druids by thousands of years, but, in his time, Aubrey was closer to the truth than anyone, including Inigo Jones, who published his book *Stonehenge Restored* in 1655. Aubrey remarks: 'Having compared his scheme with the monument itself, I found he had not dealt fairly.' Jones argued that the temples were Roman in origin and misrepresented the physical evidence to fit his theory. Such egoism was foreign to Aubrey. Moreover, were it not for his protective passion, future generations might never have had our superior opportunities to understand the mysterious stones.

To posterity, Aubrey is best known as the sociable scurrilous gossip behind *Brief Lives* (the collection of short biographies he assembled in the last decades of his long life). Aubrey did not publish his *Lives* before his own death in 1697. Nor did he publish many other manuscripts, among them *Monumenta Britannica*, *Wiltshire Antiquities* and *Naturall Historie of Wiltshire*. All are suffused by the urgent, passionate quest of a collector ever mindful of the fact that 'Death comes even for stones and names.' The manuscripts are infernally difficult to work with. Contemporary scholars are now working painstakingly towards definitive editions of Aubrey's works, which are of enormous interest to historians of antiquarianism, archaeology and science. The current selection is drawn from Andrew Clark's late nineteenth century edition of *Brief Lives*, and presents some of the scientific lives that interested Aubrey.

The one book Aubrey did publish in his lifetime was his *Miscellanies*, a collection of occult phenomena. Eighteenth century readers, priding themselves on their rationalism, often considered this a 'mad book'. It seemed to fit uneasily with Aubrey's more respectable scientific interests, and his friendships with sceptical men like Thomas Hobbes and the fellowship of the Royal Society. Magic and astrology interested him always and he went so far as to ascribe the disappointments of his own chaotic life to the unfortunate astrological aspects under which he was born in Wiltshire on 12th March 1626. It should come as no surprise then to find him beginning some of the scientific lives in this volume with a horoscope or nativity. Aubrey's life of John Dee mentions letters 'of chemistry and of magical secrets' and identifies Dee's practice of distilling eggshells as the inspiration for Ben Jonson's play *The Alchemist*.

In the best book on Aubrey's ideas, *John Aubrey and the Realm of Learning* (1975), the historian Michael Hunter argues that there was still no straightforward contrast between mechanist moderns and mystical ancients in Aubrey's time: in fact, 'he was typical in deriving scientific and magical theories and explanations from all sorts of old and new sources.' Once again, Aubrey appears essentially as a collector: one with eclectic and remarkably wide-ranging interests, who understood more sensitively than most the fragility of knowledge.

Of William Harvey he writes:

> He had made dissections of frogs, toads and a number of other animals, and had curious observations on them, which papers, together with his goods, in his lodgings

> at Whitehall, were plundered at the beginning of the Rebellion, he being for the king, and with him at Oxford; but he often said, that of all the losses he sustained, no grief was so crucifying to him as the loss of these papers, which for love or money he could never retrieve or obtain.

No one could have felt Harvey's loss as closely as Aubrey who feared constantly for the safety of his own collections. Fortunately for us they remain, for the most part, safe in the Bodleian Library.

– *Ruth Scurr, 2011*

NOTE ON THE TEXT

The work for which John Aubrey is best known, his collection of biographical information often called *Aubrey's Brief Lives*, is derived from a series of notes written at different times in large notebooks. Often, Aubrey would return after a lapse of time to a particular entry and scribble an addition to it half way down the page or in the margins. Sometimes his writing is not very clear as a result of 'the unsteadiness consequent on writing in the midst of morning sickness after a night's debauch', according to the Victorian editor of the most comprehensive printed version.

In editing this selection of Scientific Lives, the main criterion has been to produce a series of readable essays rather than reproduce the irregularities and oddities – to a modern eye – of Aubrey's chaotic working methods. As a result, spellings have been simplified and modernised, Aubrey's notes to himself about issues that needed further investiga-tion have been omitted, and several long passages in Latin have been removed rather than translated. The definition of 'scientific' has also been stretched somewhat, since there were no such people as scientists in the modern sense among the people Aubrey was writing about. But Sir Thomas Chaloner, for example, was behaving scientifically when he observed the soil while hunting in Yorkshire and wondered if it contained the useful chemical, alum. He confirmed this by tasting the local water and set up the first alum works in England. A man called Robson – we are not told his first name – although no scientist, brought a method of glass manufacture to England, and since Aubrey was fond of scabrous detail we learn that some verses about his enterprise led King James to 'beshit his briggs'.

It is odd personal details like these that fascinated Aubrey and make his writings fascinating to us. But as a Fellow of the Royal Society himself, he also understood the value of the embryonic methods of scientific research and he has given us several intimate and engaging portraits of major scientific and mathematical figures such as Isaac Barrow, Robert Boyle, Edmund Halley, William Harvey and Robert Hooke.

What Aubrey lacked in thoroughness and research methodology, he made up for with his vivid and breezy observations of human talent and human frailty, and this selection has tried to capture those qualities to the full.

Scientific Lives

THOMAS ALLEN
1542–1632

Mr Thomas Allen was born in Staffordshire. Mr Theodore Haak, a German, a Fellow of the Royal Society, was of Gloucester Hall, 1626, and knew this learned worthy old gentleman, whom he takes to have been about ninety-six years old when he died, which was about 1630. The learned (Edmund) Reynolds, who was turned Catholic by his brother the learned Dr John Reynolds, President of Corpus College, was of Gloucester Hall then too. They were both near of an age, and they died both within twelve months of each other. He was at both their funerals. Mr Allen came into the hall to commons, but Mr Reynolds had his brought to his chamber.

He says that Mr Allen was a very cheerful, facetious man, and that everybody loved his company, and every house on their Gaudy-days were wont to invite him.

His picture was drawn at the request of Dr Ralph Kettle, and hangs in the dining room of the President of Trinity College, Oxford (of which house he first was, and had his education there), by which it appears that he was a handsome sanguine man, and of an excellent habit of body.

There is mention of him in *Leicester's Commonwealth* that the great Dudley, Earl of Leicester, made use of him for casting nativities, for he was the best astrologer of his time. He has written a large and learned commentary, in folio, on the *Quadripartite* of Ptolemy, which Elias Ashmole has in MS fairly written, and I hope will one day be printed.

In those dark times astrologer, mathematician, and conjurer, were accounted the same things; and the vulgar did verily believe him to be a conjurer. He had a great many mathematical instruments and glasses in his chamber, which did also confirm the ignorant in their opinion, and his servitor (to impose on freshmen and simple people) would tell them that sometimes he should meet the spirits coming up the stairs like bees. One of our parish was of Gloucester Hall about seventy years and more since, and told me this from his servitor. Now there is to some men a great lechery in lying, and imposing on the understandings of believing people, and he thought it for his credit to serve such a master.

He was generally acquainted, and every long vacation, he rode into the country to visit his old acquaintance and patrons, to whom his great learning, mixed with much sweetness of humour, rendered him very welcome. One time being at Holm Lacy in Herefordshire, at Mr John Scudamore's (grandfather to the lord Scudamore), he happened to leave his watch in the chamber window (watches were then rarities) – the maids came in to make the bed, and hearing a thing in a case cry *Tick, Tick, Tick*, presently concluded that that was his Devil, and took it by the string with the tongues, and threw it out of the window into the moat (to drown the Devil). It so happened that the string hung on a sprig of an elder that grew out of the moat, and this confirmed them that 'twas the Devil. So the good old gentleman got his watch again.

Sir Kenelm Digby loved him much and bought his excellent library of him, which he gave to the University. I have a Stifelius' *Arithmatique* that was his, which I find he had much perused, and no doubt mastered. He was interred in Trinity College Chapel. George Bathurst BD made his funeral oration in Latin, which was printed. 'Tis pity there had not been his name on the stone over him.

Thomas Allen left the house because he would not take orders.

Queen Elizabeth sent for him to have his advice about the new star that appeared in the Swan or Cassiopeia (but I think the Swan), to which he gave his judgement very learnedly.

ISAAC BARROW
1630–77

Isaac Barrow, DD – from his father (who was born 22nd April 1600, half a year older than King Charles I), 17th May 1682. His father, Thomas Barrow, was the second son of Isaac Barrow of Spinney Abbey in the county of Cambridge, esq., who was a Justice of the Peace there above forty years. The father of Thomas never designed him for a tradesman, but he was so severe to him that he could not endure to live with him and so came to London and was apprentice to a linen-draper. He kept shop at the sign of the White Horse in Forster Lane near St Forster's Church in St Leonard's parish; and his son was christened at St John Zacharie's in Forster Lane, for at that time St Leonard's Church was pulled down to be re-edified. He was born in 1630 in October after King Charles II. Dr Isaac Barrow had the exact day and hour of his father, which may be found amongst his papers. His father set it down in his English Bible, a fair one, which they used at the king's chapel when he was in France and he could not get it again. His father travelled with King Charles II, wherever he went; he was sealer to the Lord Chancellor beyond sea, and so when he came to England. Amongst Dr Barrow's papers it may be found. He went to school, first to Mr Brookes at Charterhouse two years. His father gave to Mr Brookes £4 per annum, whereas his pay was but £2, to be careful of him; but Mr Brookes was negligent of him, which the captain of the school acquainted his father (his kinsman) and said that he would not have him stay there any longer than he did, for that he instructed him.

Afterwards to one Mr Holbitch, about four years, at Felton in Essex; from whence he was admitted of Peterhouse College in Cambridge first, and went to school a year after. Then he was admitted of Trinity College in Cambridge at thirteen years old. His mother was Anne, daughter of William Buggin of North Cray in Kent, esq. She died when her son Isaac was about four years old.

His humour when a boy and after was merry and cheerful and he was beloved wherever he came. His grandfather kept him till he was seven years old: his father was fain to force him away, for there he would have been good for nothing.

A good poet, English and Latin. He spoke eight several languages.

His father dealt in his trade to Ireland where he had a great loss, near £1,000; upon which he wrote to Mr Holbitch, a Puritan, to be pleased to take a little pains more than ordinary with him, because the times growing so bad, and such a loss then received, that he did not know how he might be able to provide for him, and so Mr Holbitch took him away from the house where he was boarded to his own house, and made him tutor to my lord Viscount Fairfax, ward to the lord Viscount Say and Seale, where he continued so long as my lord continued.

This Viscount Fairfax died a young man. This Viscount Fairfax, being a schoolboy, married a gentleman's daughter in the town there, who had but £1,000. So leaving the schools, would needs have Mr Isaac Barrow with him, and told him he would maintain him. But the lord Say was so cruel to him that he would not allow anything that 'tis thought he died for want. The £1,000 could not serve him long.

During this time old Mr Thomas Barrow was shut up at Oxford and could not hear of his son. But young Isaac's master, Holbitch, found him out in London and courted him to come to his school and that he would make him his heir. But he did not care to go to school again.

When my lord Fairfax failed[1] and that he saw he grew heavy upon him, he went to see one of his schoolfellows, one Mr Walpole, a Norfolk gent, who asked him 'What he would do?' He replied he 'knew not what to do; he could not go to his father at Oxford.' Mr Walpole then told him 'I am going to Cambridge to Trinity College and I will maintain you there:' and so he did for half a year until the surrender of Oxford; and then his father enquired after him and found him at Cambridge. And the very next day after old Mr Barrow came to Cambridge, Mr Walpole was leaving the university and (hearing nothing of Isaac's father) resolved to take Isaac along with him to his house. His father then asked him what profession he would be of, a merchant or etc.? He begged of his father to let him continue in the university. His father then asked what would maintain him. He told him £20 per annum: 'I warrant you,' said he, 'I will maintain myself with it.' His father replied, 'I'll make a shift to allow you that.' So his father then went to his tutor and acquainted him of, etc. His tutor, Dr Duport, told him that he would take nothing for his reading to him, for that he was likely to make a brave scholar, and he would help him to half a chamber for nothing. And the next news his father heard of him was that he was chosen in to the house. Dr Hill was then master of the college. He met Isaac one day and laid his hand upon his head and said, 'thou art a good boy; 'tis pity that thou art a cavalier.' He was a strong and a stout man and feared not any man. He would fight with the butchers' boys in St Nicholas' shambles, and be hard enough for any of them.

He went to travel three or four years after the king was beheaded, upon the college account. He was a candidate for the Greek professor's place, and had the consent of the university but Oliver Cromwell put in Dr Widdrington; and then he travelled.

He was abroad five years, viz. in Italy, France, Germany, Constantinople.

As he went to Constantinople, two men of war (Turkish ships) attacked the vessel; which the men that were in that engagement often testify, for he never told his father of it himself.

Upon his return, he came in a ship to Venice, which was stowed with cotton wool, and as soon as ever they came on shore the ship fell on fire, and was utterly consumed, and not a man lost, but not any goods saved – a wonderful preservation.

At Constantinople, being in company with the English merchants, there was a Rhadamontade[2] that would fight with any man and bragged of his valour, and dared any man there to try him. So no man accepting his challenge, said Isaac (not then a divine), 'Why, if none else will try you I will'; and fell upon him and chastised him handsomely that he vaunted no more amongst them.

After he had been three years beyond sea, his correspondent died, so that he had no more supply; yet he was so well beloved that he never wanted.

At Constantinople he waited on the consul Sir Thomas Bendish, who made him stay with him and kept him there a year and a half, whether he would or no.

At Constantinople, Mr Dawes (afterwards Sir Jonathan Dawes, who died Sheriff of London), a Turkey merchant, desired Mr Barrow to stay but such a time and he would return with him, but when that time came he could not go, some business stayed him. Mr Barrow could stay no longer; so Mr Dawes would have had Mr Barrow have a hundred pistoles.[3] 'No,' said Mr Barrow, 'I know not whether I shall be able to pay you.' ''Tis no matter,' said Mr Dawes. To be short, forced him to take fifty pistoles, which at his return he paid him again.

His pill (an opiate, possibly), which he was wont to take in Turkey, which was wont to do him good, but he took in preposterously at Mr Wilson's, the saddler's, near Suffolk House, where he was wont to lie and where he died, and 'twas the cause of his death. As he lay expiring in the agony of death, the standers-by could hear him say softly 'I have seen the glories of the world.'

I have heard Mr Wilson say that when he was at study, was so intent at it that when the bed was made, or so, he heeded it not nor perceived it, was so *totus in hoc*;[4] and would sometimes be going out without his hat on.

He was by no means a spruce man, but most negligent in his dress. As he was walking one day in St James's Park, his hat up, his cloak half on and half off, a gent came behind him and clapped him on the shoulder and said, 'Well, go thy ways for the veriest scholar that ever I met with.' He was a strong man, but pale as the candle he studied by.

JAMES BOVEY
1622–16…

James Bovey, esq., was the youngest son of Andrew Bovey, merchant, cash-keeper to Sir Peter Vanore, in London.

He was born in the middle of Mincing Lane, in the parish of Saint Dunstan's in the East, London, 7th May 1622, at 6 o'clock in the morning. He went to school at Mercers Chapel, under Mr Augur. At nine sent into the Low Countries; then returned, and perfected himself in the Latin and Greek. At fourteen travelled into France and Italy, Switzerland, Germany and the Low Countries. Returned into England at nineteen; he then lived with one Hoste, a banker, eight years, was his cashier eight or nine years. Then traded for himself (twenty-seven) until he was thirty-one; then married the only daughter of William de Vischer, a merchant; lived eighteen years with her, then continued single. Left off trade at thirty-two, and retired to a country life, by reason of his indisposition, the air of the city not agreeing with him. Then in these retirements he wrote *Active Philosophy* (a thing not done before), wherein are enumerated all the arts and tricks practiced in negotiation, and how they were to be balanced by counter-prudential rules.

Whilst he lived with Mr Hoste, he kept the cash of the ambassadors of Spain that were here; and of the farmers, called by them *Assentistes*, that did furnish the Spanish and Imperial armies of the Low Countries and Germany; and also many other great cashes, as of Sir Theodore Mayern, etc.; his dealing being altogether in money matters; by which means he became acquainted with the ministers of state both here and abroad.

When he was abroad, his chief employment was to observe the affairs of state and their judicatures, and to take political surveys in the countries he travelled through, more especially in relation to trade. He speaks Low-Dutch, High-Dutch, French, Italian, Spanish and Lingua Franco, and Latin besides his own.

When he retired from business he studied the Law-Merchant, and admitted himself of the Inner Temple, London, about 1660. His judgement has been taken in most of the great causes of his time in points concerning the Law-Merchant. As to his person he is about five foot high, slender, straight, hair exceeding black and curling at the end, a dark hazel eye, of a middling size, but the most sprightly that I have beheld. Brows and beard of the colour as his hair. A person of great temperance, and deep thoughts, and a working head, never idle. From his youth he had a candle burning by him all night, with pen, ink, and paper, to write down thoughts as they came into his head; that so he might not lose a thought. He was ever a great lover of Natural Philosophy. His whole life has been perplexed in lawsuits (which has made him expert in human affairs), in which he always overcame. He had many lawsuits with powerful adversaries; one lasted eighteen years. Red-haired men never had any kindness for him. He used to say: '*In rufa pelle non est animus sine felle.*'[5]

In all his travels he was never robbed.

He has one son, and one daughter who resembles him.

From fourteen he began to take notice of all prudential rules as came in his way, and wrote them down, and so continued till this day, 28th September 1680, being now in his fifty-ninth year.

For his health he never had it very well, but indifferently, always a weak stomach, which proceeded from the agitation of the brain. His diet was always a fine diet: much chicken.

He wrote a table of all the Exchanges in Europe.

He made it his business to advance the trade of England, and many men have printed his conceptions.

ROBERT BOYLE
1627–91

Mr R. Boyle, when a boy at Eton was very sickly and pale – from Dr Robert Wood, who was his schoolfellow.

The honourable Robert Boyle, esq., the fifth son of Richard Boyle, the first Earl of Cork, was born at Lismore in the county of Cork, on the 25th January 1697.

He was nursed by an Irish nurse, after the Irish manner, where they put the child into a pendulous satchel (instead of a cradle), with a slit for the child's head to peep out.

He learnt his Latin… Went to the University of Leyden.

Travelled France, Italy, Switzerland. I have often times heard him say that after he had seen the antiquities and architecture of Rome, he esteemed none anywhere else.

He speaks Latin very well, and very readily, as most men I have met with. I have heard him say that when he was young, he read over Cowper's dictionary: wherein I think he did very well, and I believe he is much beholding to him for his mastership of that language.

His father in his will, when he comes to the settlement and provision for his son Robert, thus –

Item, to my son Robert, whom I beseech God to bless with a particular blessing, I bequeath, etc.

Mr R.H., who has seen the rental, says it was £3,000 per annum: the greatest part is in Ireland.

His father left him the manor of Stallbridge in Dorset, where is a great freestone house; it was forfeited by the Earl of Castlehaven.

He is very tall (about six foot high) and straight, very temperate, and virtuous, and frugal: a bachelor; keeps a coach; sojourns with his sister, the lady Ranulagh. His greatest delight is chemistry. He has at his sister's a noble laboratory and several servants (apprentices to him) to look to it. He is charitable to ingenious men that are in want, and foreign chemists have had large proof of his bounty, for he will not spare for cost to get any rare secret. At his own costs and charges he got translated and printed the New Testament in Arabic, to send into the Mahometan countries. He has not only a high renown in England, but abroad; and when foreigners come to hither, 'tis one of their curiosities to make him a visit.

HENRY BRIGGS
1556–1630

He was first of St John's College in Cambridge. Sir Henry Savile sent for him and made him his geometry professor. He lived at Merton College in Oxford, where he made the dials at the buttresses of the east end of the chapel with a bullet for the axis.

He travelled into Scotland to commune with the honourable John Napier of Merchiston about making the logarithmical tables.

Looking one time on the map of England he observed that the two rivers, the Thames and the Avon which runs to Bath and so to Bristol, were not far distant, i.e., about three miles – see the map. He sees 'twas about twenty-five miles from Oxford; gets a horse and views it and found it to be level ground and easy to be dug. Then he considered the charge of cutting between them and the convenience of making a marriage between those rivers which would be of great consequence for cheap and safe carrying of goods between London and Bristol, and though the boats go slowly and with meanders, yet considering they go day and night they would be at their journey's end almost as soon as the wagons, which often are overthrown and liquors spilt and other goods broken. Not long after this he died and the civil wars broke out. It happened by good luck that one Mr Matthews of

Dorset had some acquaintance with this Mr Briggs and had heard him discourse on it. He was an honest simple man and had run out of his estate and this project did much run in his head. He would revive it (or else it had been lost and forgot) and went into the country to make an ill survey of it (which he printed) but with not great encouragement of the country or others. Upon the restoration of King Charles II he renewed his design and applied himself to the king and council. His Majesty espoused it more (he told me) than anyone else. In short, for want of management and his non-ability, it came to nothing and he is now dead of old age. But Sir Jonas Moore (an expert mathematician and a practical man), being sent to survey the manor of Dauntsey in Wilts (which was forfeited to the crown by Sir John Danvers his foolery), went to see these streams and distances. He told me the streams were too small unless in winter; but if some prince or the Parliament would raise money to cut through the hill at Wootton Bassett which is not very high, then there would be water enough and streams big enough. He computed the charge, which I have forgotten, but I think it was about £200,000.

THOMAS BUSHELL
1594–1674

He was one of the gentlemen that waited on the Lord Chancellor Bacon. 'Twas the fashion in those days for gentlemen to have their suits of clothes garnished with buttons. My Lord Bacon was then in disgrace, and his man Bushell having more buttons than usual on his cloak, etc., they said that his lord's breech made buttons and Bushell wore them – from whence he was called 'buttoned Bushell'.

He was only an English scholar, but had a good wit and a working and contemplative head. His lord much loved him.

His genius lay most towards natural philosophy, and particularly towards the discovery, draining and improvement of the silver mines in Cardiganshire.

He had the strangest bewitching way to draw in people (yea, discreet and wary men) into his projects that ever I heard of.

His tongue was a chain and drew in so many to be bound for him and to be engaged in his designs that he ruined a number.

Mr Goodyeere of Oxfordshire was undone by him among others.

He was master of the art of running in debt, and lived so long that his debts were forgotten, so that they were the great-grandchildren of the creditors.

He wrote a stitched treatise of mines and improving of the adits[6] to them and bellows to drive in wind, which Sir John Danvers, his acquaintance, had, and nailed it to his parlour wall at Chelsea, with some scheme, and I believe is there yet: I saw it there about ten years since.

During the time of the civil wars, he lived in Lundy Island.

In 1647 or 8, he came over into England; and when he landed at Chester, and had but one Spanish threepence, and, said he, 'I could have been contented to have begged a penny, like a poor man.' At that time he said he owed, I forget whether it was £50,000 or £60,000: but he was like Sir Kenelm Digby, if he had not fourpence, wherever he came he would find respect and credit.

After his master the Lord Chancellor died, he married, and lived at Enstone, Oxfordshire; where having some land lying on the hanging of a hill facing the south, at the foot whereof runs a fine clear stream which petrifies,[7] and where is a pleasant solitude, he spoke to his servant Jack Sydenham to get a labourer to clear some boscage which grew on the side of the hill to sit, and read or contemplate. The workman had not worked an hour before he discovers not only a rock, but a rock of an unusual figure with pendants like icicles as at Wookey Hole (Somerset), which was the occasion of making that delicate grotto and those fine walks.

Here in fine weather he would walk all night. Jack Sydenham sang rarely: so did his other servant, Mr Batty. They went very gentlemanly in clothes, and he loved them as his children.

He did not encumber himself with his wife, but here enjoyed himself thus in this paradise till the war broke out, and then retired to Lundy Isle.

He had done something (I have forgotten what) that made him obnoxious to the Parliament or Oliver Cromwell, about 1650; would have been hanged if taken; printed several letters to the Parliament, etc., dated from beyond sea, and all that time lay privately in his house at Lambeth marsh where the pointed pyramid is. In the garret there, is a long gallery, which he hung all with black and had some death's heads and bones painted.

At the end where his couch was, was in an old Gothic niche (like an old monument) painted a skeleton incumbent on a mat.

At the other end, where was his pallet-bed, was an emaciated dead man stretched out. Here he had several mortifying and divine mottos (he imitated his lord as much as he could), and out of his windows a very pleasant prospect. At night he walked in the garden and orchard. Only Mr Sydenham, and an old trusty woman, was privy to his being in England.

He died in Scotland Yard near Whitehall in 1675 or 1677.

His entertainment to Queen Henrietta Marie at Enstone was in 1636.

He was a handsome proper gentleman when I saw him at his house aforesaid in Lambeth. He was about seventy but

I should have not guessed him hardly sixty. He had a perfect healthy constitution; fresh, ruddy face; hawk-nosed, and was temperate.

As he had the art of running in debt, so sometimes he was attacked and thrown into prison; but he would extricate himself again strangely.

In the time of the civil wars his hermitage over at the rocks at Enstone were hung with black baize; his bed had black curtains, etc., but it had no bedposts but hung by four cords (covered with black baize), instead of bedposts.

When the queen mother came to Oxford to the king, she either brought (as I think) or somebody gave her an entire mummy from Egypt, a great rarity, which Her Majesty gave to Mr Bushell, but I believe long before this time the dampness of the place has spoiled it with mouldiness.

The grotto below looks just south; so that when it artificially rains upon the turning of a cock,[8] you are entertained with a rainbow. In a very little pond (no bigger than a basin), opposite to the rock, and hard by, stood (8th August 1643) a Neptune, neatly cut out in wood, holding his trident in his hand, and aiming with it at a duck which perpetually turned round with him, and a spaniel swimming after her – which was very pretty, but long since spoiled. I hear that the Earl of Rochester, in whose possession it now is, keeps it very well in order.

Mr Bushell was the greatest arts-master to run in debt (perhaps) in the world. He died £100,000 in debt. He had so delicate a way of making his projects alluring and feasible, profitable, that he drew to his baits not only rich men of no design, but also the craftiest knaves in the country, such who had cozened and undone others: e.g. Mr Goodyeere, who undid Mr Nicholas Mees's father, etc.

Ask Dr Plott (author of *Antiquities of Oxfordshire*) of the book I gave him some years since of the songs and entertainment of Mr Bushell to Queen Henrietta Marie at his rocks. If he had it not, perhaps Anthony Wood had it. Mr Edmund Wyld says that he tapped the mountain of Snowdon in Wales, which was like to have drowned all the country; and they were like to knock him and his men in the head.

Mr Thomas Bushell lay some time (perhaps years) at Captain Norton's, in the gate at Scotland Yard, where he died seven years since, about eighty years old. Buried in the little cloisters at Westminster Abbey: see the Register. Somebody put BB[9] upon the stone.

WILLIAM BUTLER
1535–1618

William Butler, physician; he was of Clare Hall in Cambridge, never took the degree of Doctor, though he was the greatest physician of his time.

The occasion of his being first taken notice of was thus: about the coming in of King James, there was a minister of [a parish] a few miles from Cambridge, that was to preach before His Majesty at Newmarket. The parson heard that the king was a great scholar, and [so he] studied so excessively that he could not sleep, so somebody gave him some opium, which had made him sleep his last, had not Dr Butler used this following remedy. He was sent for by the parson's wife. When he came and saw the parson, and asked what they had done, he told her that she was in danger to be hanged for killing her husband, and so in great choler left her. It was at that time when the cows came into the backside to be milked. He turns back, and asked whose cows those were. She said her husband's. Said he, 'Will you give one of these cows to fetch your husband to life again?' That she would, with all her heart. He then causes one presently to be killed and opened, and the parson to be taken out of his bed and put into the cow's warm belly, which after some time brought him to life, or else he had infallibly died.

There is a parallel story to this in Machiavelli's *History of Florence*, where 'tis said that one of the Medici being poisoned was put into a mule's belly, sewn up, with a place only for his head to come out.

One time King James sent for him to Newmarket, and when he was gone half way he left the messenger and turned back; so then the messenger made him ride before him.

I think he was never married. He lived in an apothecary's shop in Cambridge, John Crane, to whom he left his estate; and he in gratitude erected the monument for him, at his own charge, in the fashion he used. He was not greedy of money, except choice pieces of gold or rarities.

He would many times (I have heard say) sit among the boys at St Mary's Church in Cambridge (and just so would the famous attorney-general Noy, in Lincoln's Inn, who had many such frolics and humours).

I think he left his estate to the apothecary. He gave to the chapel of Clare Hall, a bowl, for the communion, of gold (cost, I think, £200 or £300), on which is engraved a pelican feeding her young with the blood from her breast (an emblem of the passion of the Christ), no motto, for the emblem explained it all.

He lies buried in the south side of St Mary's chancel, in Cambridge, where is a decent monument, with his body half way, and an inscription.

He was much addicted to his humours, and would suffer persons of quality to wait sometimes some hours at his door, with coaches, before he would receive them. Once, on the road from Cambridge to London, he took a fancy to a chamberlain or tapster in his inn, and took him with him, and made him his favourite, by whom only accession was to be had to him, and thus enriched him. Dr Gale, of St Paul's school, assures me that a French man came one time from London to Cambridge, purposely to see him, whom he made stay two hours for him in his gallery, and then he came out to him in an old blue gown; the French gentleman makes him two or three very low bows down to the ground; Dr Butler whips his leg over his head, and away goes into his chamber, and did not speak with him.

He kept an old maid whose name was Nell. Dr Butler would many times go to the tavern, but drink by himself. About nine or ten at night old Nell comes for him with a candle and a lantern, and says, 'Come you home, you drunken beast.' By and by Nell would stumble; then her master calls her 'drunken beast'; and so they did 'drunken beast' another all the way till they came home.

A serving man brought his master's water to Dr Butler, being then in his study (with turned bars[10]) but would not be spoken with. After much fruitless importunity, the man told the doctor he was resolved he should see his master's water; he would not be turned away – threw it on the doctor's head.

The humour pleased the doctor and he went to the gentleman and cured him.

A gentleman, lying a-dying, sent his servant with his horse for the doctor. The horse being exceedingly dry, ducks down his head strongly into the water, and plucks down the doctor over his head, who was plunged in the water over head and ears. The doctor was maddened, and would return home. The man swore he should not; drew his sword, and gave him ever and anon (when he would return) a little prick, and so drove him before him.

Some instances of Dr Butler's cures.

The doctor lying at the Savoy in London, next the water side, where was a balcony looked into the Thames, a patient came to him that was grievously tormented with an ague. The doctor orders a boat to be in readiness under his window, and discoursed with the patient (a gentleman) in the balcony, when on a signal given, two or three lusty fellows came behind the gentleman and threw him a matter of twenty feet into the Thames. This surprise absolutely cured him.

A gentleman with a red, ugly, pimpled face came to him for a cure. Said the doctor, 'I must hang you.' So presently he had a device made ready to hang him from a beam in the room; and when he was even almost dead, he cuts the veins that fed these pimples, and let out the black ugly blood, and cured him.

Another time one came to him for the cure of a cancer (or ulcer) in the bowels. Said the doctor, 'Can ye shit?' 'Yes,' said the patient. So the doctor ordered a basin for him to shit, and when he had so done the doctor ordered him to eat it up. This did the cure.

That he was chemical[11] I know by this token that his maid came running in to him one time, like a slut and a fury, with her hair about her ears, and cries, 'Butler! come and look to your devils yourself, and you will: the stills are all blown up!' She tended them, and it seems gave too great a heat.

THOMAS CHALONER
1595–1661

Thomas Chaloner, esq., [bred up in Oxon], was the [third] son of Dr [Thomas] Chaloner.

He was a well-bred gentleman, and of very good natural parts, and of an agreeable humour. He had the accomplishments of studies at home, and travels in France, Italy and Germany.

Riding a hunting in Yorkshire (where the alum works now are), on a common, he took notice of the soil and herbage, and tasted the water, and found it to be like that where he had seen the alum works in Germany. Whereupon he got a patent of the king (Charles I) for an alum work (which was the first that ever was in England), which was worth to him £2,000 per annum, or better: but in the time of Charles I some courtiers did think the profit too much for him, and prevailed so with the king, that, notwithstanding the patent aforesaid, he granted a moiety[12], or more, to another (a courtier), which was the reason that made Mr Chaloner so interest himself for the Parliament-cause, and, in revenge, to be one of the king's judges.

He was as far from a puritan as the East from the West. He was of the natural religion, and of Henry Martyn's gang, and one who loved to enjoy the pleasures of this life. He was (they say) a good scholar, but he wrote nothing that I hear of, only an anonymous pamphlet, *An account of the Discovery of Moyses's Tombe*; which was written very wittily. It was about 1652. It did set the wits of all the Rabbis of the Assembly then to work, and 'twas a pretty while before the sham was detected.

He had a trick sometimes to go into Westminster Hall in morning in term time, and tell some strange story (sham), and would come thither again about eleven or twelve to have the pleasure to hear how it spread: and sometimes it would be altered, with additions, he could scarce know it to be his own. He was neither proud nor covetous, nor a hypocrite: not apt to do injustice, but apt to revenge.

After the restoration of King Charles II, he kept the castle at the Isle of Man, where he had a pretty wench that was his concubine; where when news was brought to him that there were some come to the castle to demand it for His Majesty, he spoke to his girl to make him a posset, into which he put, out of a paper he had, some poison, which did, in a very short time, make him fall a vomiting exceedingly; and after some time vomited nothing but blood. His retchings were so violent that the standers by were much grieved to behold it. Within three hours he died. The demandants of the castle came and saw him dead; he was swollen so extremely that they could not see any eye he had, and no more of his nose than the tip of it, which showed like a wart, and his coddes[13] were swollen as big as one's head. This account I had from George Estcourt, DD, whose brother-in-law was one of those that saw him.

EDWARD DAVENANT
16… –1679/80

Edward Davenant was the eldest son of [Edward] Davenant, merchant of London, who was elder brother to the right reverend father in God, the learned John Davenant, Bishop of Sarum.

I will first speak of the father, for he was a rare man in his time, and deserves to be remembered. He was of a healthy complexion, rose at 4 or 5 in the morning, so that he followed his studies till 6 or 7, the time that other merchants go about their business; so that, stealing so much and so quiet time in the morning, he studied as much as most men. He understood Greek and Latin perfectly, and was a better Grecian than the bishop. He writ a rare Greek character as ever I saw. He was a great mathematician, and understood as much of it as was known in his time. Dr Davenant, his son, has excellent notes of his father's, in mathematics, as also in Greek, and 'twas no small advantage [to] him to have such a learned father to imbue arithmetical knowledge into him when a boy, night times when he came from school (Merchant Taylors'). He understood trade very well, was a sober and good manager, but the winds and seas crossed him. He had so great losses that he broke, but his creditors knowing it was no fault of his, and also that he was a person of great virtue and justice, used not extremity towards him; but I think gave him more credit, so that he went into Ireland, and did set up a fishery for pilchards at Wythy Island, in Ireland, where he got £10,000; satisfied and paid his creditors; and over and above left a good estate to his son. His picture bespeaks him to be a man of judgement, and parts, and gravity extraordinary. He slipped coming down the stone stairs at the palace at Sarum, which bruise caused his death. He lies buried in the south aisle of the choir in Sarum Cathedral behind the bishop's stall.

Dr Edward Davenant was born at his father's house in Croydon in Surrey (the farthest handsome great house on the left hand as you ride to Banstead Downs). I have heard him say, he thanked God his father did not know the hour of his birth; for that it would have tempted him to have studied astrology, for which he had no esteem at all.

He went to school at Merchant Taylors' school, from thence to Queen's College in Cambridge, of which house his uncle,

John Davenant (afterwards bishop of Sarum), was head, where he was fellow.

When his uncle was preferred to the church of Sarum, he made his nephew treasurer of the church, which is the best dignity, and gave him the vicarage of Gillingham in Dorset, and then Paulshot parsonage, near Devizes, which last in the late troubles he resigned to his wife's brother [William] Grove.

He was to his dying day of great diligence in study, well versed in all kinds of learning, but his genius did most strongly incline him to the mathematics, wherein he has written (in a hand as legible as print) MSS in quarto a foot high at least. I have often heard him say (jestingly) that he would have a man knocked in the head that should write anything in mathematics that had been written of before. I have heard Sir Christopher Wren say that he does believe he was the best mathematician in the world about thirty or thirty-five years ago. But being a divine he was unwilling to print, because the world should not know how he had spent the greatest part of his time.

He very rarely went any further than the church, which is hard by his house. His wife was a very discreet and excellent housewife, that he troubled himself about no mundane affairs, and 'tis a private place, that he was but little diverted with visits.

I have written to his executor, that we may have the honour and favour to conserve his MSS in the Library of the Royal Society, and to print what is fit. I hope I shall obtain my desire. And the Bishop of Exeter ([Thomas] Lamplugh) married the doctor's second daughter Katherine, and he was tutor to Sir Joseph Williamson, our President. He had a noble library, which was the aggregate of his father's, the bishop's, and his own.

He was of middling stature, something spare; and weak, feeble legs; he had sometimes the gout; was of great temperance, he always drank his beer at meals with a toast, winter and summer, and said it made the beer the better.

He was not only a man of vast learning, but of great goodness and charity; the parish and all his friends will have a great loss in him. He took no use for money upon bond. He was my singular good friend, and to whom I have been more beholding then to any one beside: for I borrowed £500 of him for a year and a half, and I could not fasten any interest on him.

He was very ready to teach and instruct. He did me the favour to inform me first in algebra. His daughters were algebrists.

His most familiar learned acquaintance was Lancelot Morehouse, parson of Pertwood. I remember when I was a young Oxford scholar, that he could not endure to hear of the *New* (Cartesian, or etc.) *Philosophy*; 'For,' said he, 'if a new philosophy is brought in, a new divinity will shortly follow' (or 'come next'); and he was right.

He died at his house at Gillingham aforesaid, where he and his predecessor, Dr [John] Jessop, had been vicars and lies buried in the chancel there. He died on 9th March 1679/80, was buried the 31st of the same month.

He was heir to his uncle, John Davenant, bishop of Sarum. When Bishop Coldwell came to this bishopric, he did let long leases, which were but newly expired when Bishop Davenant came to this see; so that there tumbled into his coffers vast sums. His predecessor, Dr Tounson, married his sister, continued in the see but a little while, and left several children unprovided for, so the king or rather duke of Bucks gave Bishop Davenant the bishopric out of pure charity. Sir Anthony Weldon says (in his *Court of King James*), 'twas the only bishopric that he disposed of without simony, all others being made merchandise of for the advancement of his kindred. Bishop Davenant being invested, married all his nieces to clergymen, so he was at no expense for their preferment. He granted to his nephew (this Dr) the lease of the great manor of Poterne, worth about £1,000 per annum; made him treasurer of the church of Sarum, of which the corps is the parsonage of Calne, which was esteemed to be of the like value. He made several purchases, all which he left him; insomuch as the churchmen of Sarum say, that he gained more by this church than ever any man did by the church since the Reformation, and take it very unkindly that, at his death, he left nothing (or but £50) to that church which was the source of his estate. How it happened I know not, or how he might be worked on in his old age, but I have heard several years since, he had set down £500 in will for the Cathedral Church of Sarum.

He had six sons and four daughters. There was a good school at Gillingham: at winter nights he taught his sons arithmetic and geometry; his two elder daughters, especially Mrs Ettrick, was a notable algebrist.

He had an excellent way of improving his children's memories, which was thus: he would make one of them read a chapter or etc., and then they were (*sur le champ*[14]) to repeat what they remembered, which did exceedingly profit them; and so for sermons, he did not let them write notes (which jaded their memory), but gave an account *viva voce*.[15] When his eldest son, John, came to Winton school (where the boys were enjoined to write sermon notes) he had not wrote; the master asked him for his notes – he had none, but said, 'If I do not give you as good an account of it as they that do, I am much mistaken.'

JOHN DEE
1527–1608

His father was Roland Dee, a Radnorshire gentleman, and he was descended from Rees, Prince of South Wales.

My great-grandfather William Aubrey and he were cousins, and intimate acquaintances. Mr Ashmole has letters between them, under their own hands, viz. one of Dr W.A. to him (ingeniously and learnedly written), touching the *Sovereignty of the Sea*, of which J.D. wrote a book which he dedicated to Queen Elizabeth and desired my great-grandfather's advice upon. Dr A.'s country house was at Kew, and J. Dee lived at Mortlake, not a mile distant. I have heard my grandmother say they were often together.

Arthur Dee, MD, his son, lived and practised at Norwich, an intimate friend of Sir Thomas Browne, MD, who told me that Sir William Boswell, the Dutch ambassador, had all John Dee's MSS. He lived then somewhere in Kent.

The MSS in the Bodleian Library of Dr Gwyn, wherein are several letters between him and John Dee, and Dr Davies, of chemistry and of magical secrets, which my worthy friend Mr Meredith Lloyd has seen and read: and he tells me that he has been told that Dr Barlow gave it to the Prince of Tuscany.

Meredith Lloyd says that John Dee's printed book of spirits, is not above the third part of what was written, which were in Sir Robert Cotton's library: many whereof were much perished by being buried, and Sir Robert Cotton bought the field to dig after it.

He told me of John Dee, etc., conjuring at a pool in Brecknockshire, and that they found a wedge of gold; and that they were troubled and indicted as conjurors at the assizes;

that a mighty storm and tempest was raised in harvest time, the country people had not known the like.

His picture in a wooden cut is at the end of Billingsley's *Euclid*, but Mr Elias Ashmole has a very good painted copy of him from his son Arthur. He had a very fair, clear complexion (as Sir Henry Savile); a long beard as white as milk. A very handsome man.

Old goodwife Faldo (a native of Mortlake in Surrey) did know Dr Dee and told me he died at his house in Mortlake, next to the house where the tapestry hangings are made, viz. west of that house; and that he died about sixty plus eight or nine years since (January 1672) and lies buried in the chancel, and had a stone (marble) upon him.

Her mother tended him in his sickness. She told me that he did entertain the Polish ambassador at his house in Mortlake, and died not long after; and that he showed the eclipse with a dark room to the said ambassador. She believes that he was eighty years old when he died. She said, he kept a great many stills going. That he laid the storm by magic. That the children dreaded him because he was accounted a conjuror. He recovered the basket of clothes stolen, when she and his daughter (both girls) were negligent: she knew this.

He is buried (upon the matter) in the middle of the chancel, a little toward the south side. She said, he lies buried in the chancel between Mr Holt and Mr Miles, both servants to Queen Elizabeth, and both have brass inscriptions on their marble, and that there was on him a marble, but without any inscription, which marble is removed; on which old marble is sign of two or three brass pins. A daughter of his married to a flax dresser, in Southwark.

He died within a year, if not shortly, after the king of

Denmark was here: see Sir Richard Baker's *Chronicle* and Captain Wharton's *Almanac*.

He built the gallery in the church at Mortlake. Goody Faldo's father was the carpenter that worked it.

A stone was on his grave, which is since removed. At the upper end of the chancel then were steps, which in Oliver's days were laid plain by the minister, and then 'twas removed.

The children when they played in the church would run to Dr Dee's gravestone. She told me that he forewarned Queen Elizabeth of Dr Lopez's attempt against her (the doctor bewrayed,[16] beshit himself).

He used to distil eggshells, and 'twas from hence that Ben Jonson had his hint of the alchemist, whom he meant.

He was a great peacemaker; if any of the neighbours fell out, he would never let them alone till he had made them friends.

He was tall and slender. He wore a gown like an artist's gown, with hanging sleeves, and a slit.

A mighty good man he was.

He was sent ambassador for Queen Elizabeth into Poland.

SIR KENELM DIGBY
1603–65

Sir Kenelm Digby, knight. He was born at Gayhurst, Bucks, on 11th June: see Ben Jonson:

Witness thy actions done at Scanderoon
Upon thy birthday, the eleventh of June.

(Memorandum – in the first impression in an octavo book it is thus, but in the folio 'tis *my* instead of *thy*.) Mr Elias Ashmole assures me, from two or three nativities by Dr Richard Napier, that Ben Jonson was mistaken and did it for the rhyme's sake.

He was the eldest son of Sir Everard Digby, who was accounted the handsomest gentleman in England. Sir Everard suffered as a traitor in the gunpowder-treason, but King James restored his estate to his son and heir. His second son was Sir John Digby, as valiant a gentleman and as good a swordsman as was in England, who died (or was killed) in the king's cause at Bridgewater, about 1644. It happened in 1647 that a grave was opened next to Sir John Digby's (who was buried in summer time, it seems), and the flowers on his coffin were found fresh, as I heard Mr Harcourt (that was executed) attest that very year. Sir John died a bachelor.

Sir Kenelm Digby was held to be the most accomplished cavalier of his time. He went to Gloucester Hall in Oxon in 1618. The learned Mr Thomas Allen (then of that house) was wont to say that he was the *Mirandula* of his age. He did not wear a gown there, as I have heard my cousin Whitney say.

There was a great friendship between him and Mr Thomas Allen; whether he was his scholar I know not. Mr Allen was one of the learnedest men of this nation in his time, and a great collector of good books, which collection Sir Kenelm bought (Mr Allen enjoying the use of them for his life) to give to the Bodleian Library, after Mr Allen's decease, where they now are.

He was a great traveller, and understood ten or twelve different languages. He was not only master of a good and graceful judicious style, but he also wrote a delicate hand, both fast-hand and Roman. I have seen letters of his writing to the father of this Earl of Pembroke, who much respected him.

He was such a goodly handsome person, gigantic and great voice, and had so graceful elocution and noble address, etc., that had he been dropped out of the clouds in any part of the world, he would have made himself respected.

He was envoyé from Henrietta Maria (then queen mother) to Pope Innocent X where at first he was mightily admired; but after some time he grew, and hectored with His Holiness and gave him the lie. The Pope said he was mad.

He was well versed in all kinds of learning. And he had also this virtue, that no man knew 'better how to abound and to be abased', and either was indifferent to him. No man became grandeur better; sometimes again he would live only with a lackey, and horse with a foot-cloth.[17] He was very generous, and liberal to deserving persons.

When Abraham Cowley was but thirteen years old, he dedicated to him a comedy, called *Love's Riddle*, and concludes in his epistle – 'The Birch that whip't him then would prove a Bay.' Sir K. was very kind to him.

Much against his mother's consent, he married that celebrated beauty and courtesan, Mrs Venetia Stanley, whom Richard Earl of Dorset kept as his concubine, had children by her and settled on her an annuity of £500 per annum; which after Sir K.D. married was unpaid by the earl; and for which annuity Sir Kenelm sued the earl, after marriage, and recovered it. He would say that a handsome lusty man who was discreet might make a virtuous wife out of a brothel-house. This lady carried herself blamelessly, yet (they say) he was jealous of her. She died suddenly, and hard-hearted women would censure him severely.

After her death, to avoid envy and scandal, he retired into Gresham College at London, where he diverted himself with his chemistry and the professors' good conversation.

He wore there a long mourning cloak, a high crowned hat, his beard unshorn, looked like a hermit, as signs of sorrow for his beloved wife, to whose memory he erected a sumptuous monument, now quite destroyed by the great conflagration.

He stayed at the college two or three years.

The fair houses in Holborn, between King's Street and Southampton Street (which break off the continuance of them), were, about 1633, built by Sir Kenelm; where he lived before the civil wars. Since the restoration of Charles II he lived in the last fair house westward in the north portico of Covent Garden, where my lord Denzil Hollis lived since. He had a laboratory there. I think he died in this house.

He was, 1643, prisoner for the king (Charles I) at Winchester House, where he practised chemistry, and wrote his book of Bodies and Soul, which he dedicated to his eldest son, Kenelm, who was slain (as I take it), in the Earl of Holland's rising.

In the time of Charles I he received the sacrament in the chapel at Whitehall, and professed the Protestant religion, which gave great scandal to the Roman Catholics; but afterwards he *looked back*.

He was a person of very extraordinary strength. I remember one at Sherborne (relating to the Earl of Bristol) protested to us, that as he, being a middling man, being sat in a chair, Sir Kenelm took him up, chair and all, with one arm.

He was of undaunted courage, yet not apt in the least to give offence. His conversation was both ingenious and innocent.

As for that great action of his at Scanderoon, see the Turkish History. Sir Edward Stradling, of Glamorganshire, was then his vice-admiral, at whose house is an excellent picture of his, as he was at that time: by him is drawn an armillary sphere broken, and underneath is written IMPAVIDUM FERIENT[18] (Horace).

There is in print in French, and also in English (translated by Mr James Howell), a speech that he made at a philosophical assembly at Montpellier, *Of the Sympathique Powder*. He made a speech at the beginning of the meeting of the Royal Society *Of the Vegetation of Plants*.

He was born to £3,000 per annum. His ancient seat (I think) is Gayhurst in Buckinghamshire. He had a fair estate also in Rutlandshire. What by reason of the civil wars, and his generous mind, he contracted great debts, and I know not how (there being a great falling out between him and his *then only son*, John) he settled his estate upon Cornwallis, a subtle solicitor, and also a member of the House of Commons, who did put Mr John Digby to much charge in law.

Mr J.D. had a good estate of his own, and lived handsomely then at what time I went to him two or three times in order to your *Oxford Antiquities*;[19] and he then brought me a great book, as big as the biggest church Bible I ever saw, and the richliest bound, bossed with silver, engraved with escutcheons and crest (an ostrich); it was a curious vellum. It was the history of the family of the Digbys, which Sir Kenelm either did, or ordered to be done. There was inserted all that was to be found anywhere relating to them, out of records of the Tower, rolls, etc. All ancient church monuments were most exquisitely limned by some rare artist. He told me that the compiling of it did cost his father £1,000. Sir John Fortescue said he did believe 'twas more. When Mr John Digby did me the favour to show me this rare MS, 'This book', said he, 'is all that I have left me of all the estate that was my father's!' He was almost as tall and as big as his father: he had something of the sweetness of his mother's face. He was bred by the Jesuits, and was a good scholar.

Sir John Hoskyns informs me that Sir Kenelm Digby did translate Petronius Arbiter into English.

'Tis said there was more hurt done by the cavaliers (during their garrison) by way of embezzling and cutting off chains of books, than there was since. He was a lover of learning, and had he not taken this special care, that noble library had been utterly destroyed – for there were ignorant senators enough who would have been contented to have had it so. This I do assure you from an ocular witness, E.W., esq.

LEONARD DIGGES

15...–71

Leonard Digges, esquire, of Wotton in Kent wrote a thin folio called *Pantometria*, printed 15[71]. At the end he discourses of regular solids, and I have heard the learned Dr John Pell say it is done admirably well. In the preface he speaks of cutting glasses in such a particular manner that he would discern pieces of money a mile off; and this he says he sets down the rather because several are yet living that have seen him do it.

A quarto '*Tectonicon* briefly showing the exact measuring and speedy reckoning all manner of land, squares, timber, stone, steeples, pillars, globes etc., for declaring the perfect making and large use of the carpenter's ruler, containing a quadrant geometrical, comprehending also the rare use of the square, and in the end a little treatise opening the composition and appliance of an instrument called "The Profitable Staffe", with other things pleasant and necessary, most conducible for surveyors, land-meaters, joiners, carpenters, and masons: published by Leonard Digges, gentleman.'

L.D. to the Reader:

Although many have put forth sufficient and certain rules to measure all manner of superficies, etc. yet in that the art of numbering has been required, chiefly those rules hid and as it were locked up strange tongues, they do profit or have furthered very little, for the most part, nothing at all, the land-meater, carpenter, mason, wanting the aforesaid. For their sakes I am here provoked not to hide but to open the talent I have received, to publish in this our tongue very shortly if God give life a volume containing the flowers of the sciences mathematical largely applied to our outward practice profitably pleasant to all manner men. Here my advice shall be to those artificers, that will profit in this or any of my books now published, or that hereafter shall be, first confusedly to read them through, then with more judgement, read at the third reading wittily to practise. So, few things shall be unknown. Note, oft diligent reading joined with ingenious practice causes profitable labour. Thus most heartily farewell, loving reader, to whom I wish myself present to further thy desire and practice in these.

The method that carpenters etc. used before this book was published was very erroneous, as he declares.

THOMAS DIGGES
15..-95

Mr Thomas Digges: he wrote a book in quarto entitled: '*Stratioticos*, compendiously teaching the science of numbers as well in fractions as integers, and so much of the rules and equations algebraical and art of numbers cossicall as are requisite for the profession of a soldier; together with the modern military discipline, offices, laws and orders in every well-governed camp and army inviolably to be observed.'

First published by him, 1579, and dedicated 'unto the right honourable Robert, Earle of Leicester.' The second edition 1590.

He was muster-master general of all her majesty's forces in the Low Countries.

At the end of this book (the last paragraph) speaking of 'engines and inventions not usual to be thought on and had in readiness':

> Of these and many more important matters military, I shall have occasion at large to dilate in my treatise of great artillery and pyrotechnic, whose publication I have for divers due respects hitherto differed.

He was the only son of the learned Leonard Digges, esq., of whom he speaks in the preface to his *Stratioticos*.

In this family have been four learned men in an uninterrupted descent – *scilicet*, two eminent mathematicians (Leonard and Thomas), Sir Dudley Digges, Master of the Rolls, and his son Dudley, fellow of All Souls College, Oxford.

HENRY GELLIBRAND
1597–1637

Henry Gellibrand was born in London. He was of Trinity College in Oxford. Dr Potter and Dr [William] Hobbes knew him. Dr Hannibal Potter was his tutor, and preached his funeral sermon in London. They told me that he was good for little a great while, till at last it happened accidentally, that he heard a geometric lecture. He was so taken with it, that immediately he fell to studying it, and quickly made great progress in it. The fine dial over the College Library is of his own doing. He was astronomy professor in Gresham College. He being one time in the country, showed the tricks of drawing what card you touched, which was by combination with his confederate, who had a string that was tied to his leg, and the leg of the other, by which his confederate gave him notice by touch; but by this trick, he was reported to be a conjuror.

JONATHAN GODDARD
1617–74

Jonathan Godard [sic], MD, born at Greenwich (or Rochester, where his father commonly lived; but, to my best remembrance, he told me at the former). His father was a ship-carpenter.

He was of Magdalen Hall, Oxford. He was one of the College of Physicians, in London; Warden of Merton College, Oxford; physician to Oliver Cromwell, Protector; went with him into Ireland. [Possibly] also sent to him into Scotland, when he was so dangerously ill there of a kind of calenture or high fever, which made him mad that he pistolled one or two of his commanders that came to visit him in his delirious rage.

Praelector in medicine at Gresham College; where he lived, and had his laboratory for chemistry. He was an admirable chemist.

He had three or four medicines wherewith he did all his cures: a great ingredient was *Radix Serpentaria*.

He intended to have left his library and papers to the Royal Society, had he made his will, and had not died so suddenly. So that his books (a good collection) are fallen into the hands of a sister's son, a scholar in Caius College, Cambridge. But his papers are in the hands of Sir John Bankes, Fellow of the Royal Society. There were his lectures at Chirurgions' hall; and two manuscripts in quarto, thick volumes, ready for the press, one was a kind of Pharmacopoeia (his nephew has this). 'Tis possible his rare universal medicines aforesaid might be retrieved amongst his papers. My Lord Brounker has the recipe but will not impart it.

He was a fellow of the Royal Society, and a zealous member for the improvement of natural knowledge amongst them. They made him their drudge, for when any curious experiment was to be done they would lay the task on him.

He loved wine and was most curious in his wines, was hospitable, but drank not to excess, but it happened that coming from his club at the Crown tavern in Bloomsbury, a foot, 11 at night, he fell down dead of an apoplexy in Cheapside, at Wood-street end, 24th March 1674–5, aged fifty-six. Tomb in the church of Great St Helen, London.

EDMUND GUNTER
1581–1626

Captain Ralph Gretorex, mathematical instrument maker in London, said that he [Gunter] was the first that brought mathematical instruments to perfection. His book of the quadrant, sector, and cross-staff did open men's understandings and made young men in love with that study. Before, the mathematical sciences were locked up in the Greek and Latin tongues and so lay untouched, kept safe in some libraries. After Mr Gunter published his book, these sciences sprang up amain, more and more to that height it is at now (1690).

When he was a student at Christ Church, it fell to his lot to preach the Passion sermon, which some old divines that I knew did hear, but they said that 'twas said of him then in the university that our Saviour never suffered so much since his passion as in that sermon, it was such a lamentable one: *Non omnia possumus omnes.*[20]

The world is much beholding to him for what he has done well.

Gunter is originally a Brecknockshire family, of Tregunter. They came thither under the conduct of Sir Bernard Newmarch when he made the conquest of that county.

EDMUND HALLEY
1656–1742

Mr Edmund Halley, astronomer, born 29th October 1656, London – this nativity I had from Mr Halley himself.

Mr Edmund Halley, Master of Arts, the eldest son of Mr Halley, a soap-boiler, a wealthy citizen of the city of London; of the Halleys of Derbyshire, a good family.

He was born in Shoreditch parish, at a place called Haggerston, the backside of Hogsdon.

At nine years old, his father's apprentice taught him to write, and arithmetic. He went to Paul's school to Dr Gale; while he was there he was very perfect in the celestial globes, insomuch that I heard Mr Moxon (the globe-maker) say that if a star were misplaced in the globe, he would presently find it.

He studied geometry, and at sixteen could make a dial, and then, he said, thought himself a brave fellow.

At sixteen went to Queen's College in Oxford, well versed in Latin, Greek and Hebrew: where, at the age of nineteen, he solved this useful problem in astronomy, never done before, viz. 'from three distances given from the sun, and angles between, to find the orb', for which his name will be ever famous.

Left Oxford, and lived at London with his father till 1676; at which time he got leave, and a *viaticum* of his father, to go to the island of Saint Helena, purely upon the account of advancement in astronomy, to make the globe of the southern hemisphere right, which before was very erroneous, as being done only after the observations of ignorant seamen. At his return, he presented his planisphere, with a short description, to His Majesty who was very well pleased with it; but received nothing but praise.

I have often heard him say that if His Majesty would be but only at the charge of sending out a ship, he would take the longitude and latitude, right ascensions and declinations of the southern fixed stars.

In 1678, he added a spectacle glass to the shadow-vane of the lesser arch of the sea quadrant (or back-staff); which is of great use, for that spot of light will be manifest when you cannot see any shadow.

Cardinal d'Estrée caressed him and sent him to his brother the admiral with a letter of recommendation. He has contracted an acquaintance and friendship with all the eminentest mathematicians of France and Italy, and holds a correspondence with them.

He returned into England, 24th January 1682.

THOMAS HARIOT
1560–1621

There is a place in Kent called Hariot's-ham, now my lord Wotton's; and in Worcestershire in the parish of Droitwich is a fine seat called Hariots, late the seat of Chief Baron Wyld.

Sir Robert Moray declared at the Royal Society – 'twas when the comet appeared before the Dutch war – that Sir Francis had heard Mr Hariot say that he had seen nine comets, and had predicted seven of them, but did not tell them how. 'Tis very strange: *excogitent astronomi*.[21]

Mr Hariot went with Sir Walter Raleigh into Virginia, and has written the Description of Virginia which is printed.

Dr Pell tells me that he finds amongst his papers (which are now, 1684, in Dr Busby's hands), an alphabet that he had contrived for the American language, like devils.[22]

When Henry Percy, ninth Earl of Northumberland, and Sir Walter Raleigh were both prisoners in the Tower, they grew acquainted, and Sir Walter Raleigh recommended Mr Hariot to him, and the earl settled an annuity of £200 a year on him for his life, which he enjoyed.

But to Hues (who wrote *De Usu Globorum*[23]) and to Mr Warner he gave an annuity but of £60 per annum.

These three were usually called 'the Earl of Northumberland's Three Magi'. They had a table at the earl's charge, and the earl himself had them to converse with, singly or together.

He was a great acquaintance of Master Ailesbury, to whom Dr Corbet sent a letter in verse, 9th December 1618, when the great blazing star appeared,

Now for the peace of Gods and men advise,
(Thou that hast wherwithall to make us wise),

Thine owne rich studies a deep Harriot's mine,
In which there is no drosse but all refine.

The bishop of Salisbury (Seth Ward) told me that one Mr Hagger (a countryman of his), a gentleman and good mathematician, was well acquainted with Mr Thomas Hariot, and was wont to say, that he did not like (or valued not) the old story of the creation of the world. He could not believe the old position; he would say *ex nihilo nihil fit.*[24] But said Mr Hagger, a *nihilum*[25] killed him at last: for in the top of his nose came a red speck (exceedingly small), which grew bigger and bigger, and at last killed him. I suppose it was that which the surgeons call a *noli me tangere.*[26] Mr Hariot died of an ulcer in his lip or tongue – see Dr Read's *Surgery*, where he mentions him as his patient, in the treatise of ulcers (or cancers).

He made a philosophical theology, wherein he cast off the Old Testament, and then the new one would (consequently) have no foundation. He was a Deist. His doctrine he taught to Sir Walter Raleigh, Henry, Earl of Northumberland, and some others. The divines of those times looked on his manner of death as a judgement upon him for nullifying the Scripture.

WILLIAM HARVEY
1578–1657

William Harvey, MD, born at Folkestone in Kent: born at the house which is now the post-house, a fair stone-built house, which he gave to Caius College in Cambridge, with some lands there. His brother Eliab would have given any money or exchange for it, because 'twas his father's, and they all born there; but the doctor (truly) thought his memory would better be preserved this way, for his brother has left noble seats, and about £3,000 per annum, at least.

Dr Harvey added (or was very bountiful in contributing to) a noble building of Roman architecture (of rustic work, with Corinthian pilasters) at the Physicians' College, aforesaid, viz. a great parlour for the fellows to meet in, below; and a library, above.

All these remembrances and building was destroyed by the general fire.[27]

He was always very contemplative, and the first that I hear of that was curious about anatomy in England. He had made dissections of frogs, toads and a number of other animals, and had curious observations on them, which papers, together with his goods, in his lodgings at Whitehall, were plundered at the beginning of the Rebellion, he being for the king, and with him at Oxford; but he often said, that of all the losses he sustained, no grief was so crucifying to him as the loss of these papers, which for love or money he could never retrieve or obtain. When Charles I by reason of the tumults left London, he attended him, and was at the fight of Edgehill with him; and during the fight, the prince and Duke of York were committed to his care: he told me that he withdrew with them under a hedge, and took out of his pocket a book and read; but he had not read very long before a bullet of a great gun grazed on the ground near him, which made him remove his station.

He told me that Sir Adrian Scrope was dangerously wounded there, and left for dead amongst the dead men, stripped; which happened to be the saving of his life. It was cold, clear weather, and a frost that night; which staunched his bleeding, and about midnight, or some hours after his hurt, he awaked, and was fain to draw a dead body upon his for warmth's sake.

After Oxford was surrendered, which was 24th July 1646, he came to London, and lived with his brother Eliab a rich merchant in London, opposite to St Lawrence Poultry Church, where was then a high leaden steeple (there were but two, viz. this and St Dunstan's in the east) and at his brother's country house at Roehampton.

His brother Eliab brought, about 1654, Cockaine House, now (1680) the Excise Office, a noble house, where the doctor was wont to contemplate on the leads of the house, and had his several stations, in regard of the sun, or wind.

He did delight to be in the dark, and told me he could then best contemplate. He had a house heretofore at Combe, in Surrey, a good air and prospect, where he had caves made in the earth, in which in summer time he delighted to meditate. He was pretty well versed in the mathematics, and had made himself master of Mr Oughtred's *Clavis Math.*[28] in his old age; and I have seen him perusing it, and working problems not long before he died, and that book was always in his meditating apartment.

His chamber was that room that is now the office of Elias Ashmole, esq.; where he died, being taken with the dead palsy, which took away his speech.

He died worth £20,000 which he left to his brother Eliab. In his will he left his old friend Mr Thomas Hobbes £10 as a token of his love.

He was wont to say that man was but a mischievous baboon.

He would say, that we Europeans knew not how to order or govern our women, and that the Turks were the only people used them wisely.

He was far from bigotry.

He had been physician to Lord Chancellor Bacon, whom he esteemed much for his wit and style, but would not allow him to be a great philosopher. 'He writes philosophy like a Lord Chancellor,' he said to me, speaking in derision; 'I have cured him.' About 1649 he travelled again into Italy, Dr George (now Sir George) Ent, then accompanying him.

At Oxford, he grew acquainted with Dr Charles Scarborough, then a young physician (since by King Charles II knighted), in whose conversation he much delighted; and whereas before, he marched up and down with the army, he took to him and made him lie in his chamber, and said to him, 'Prithee leave off thy gunning, and stay here; I will bring thee into practice.'

I remember he kept a pretty young wench to wait on him, which I guess he made use of for warmth's sake as King David did, and took care of her in his will, as also of his man servant.

For twenty years before he died he took no manner of care about his worldly concerns, but his brother Eliab, who was a very wise and prudent manager, ordered all not only faithfully, but better than he could have done himself.

He was, as all the rest of the brothers, very choleric; and in his young days wore a dagger (as the fashion then was, nay I remember my old schoolmaster, old Mr Latimer, at seventy, wore a dudgeon, with a knife, and bodkin, as also my old grandfather Lyte, and alderman Whitson of Bristowe, which I suppose was the common fashion in their young days), but this doctor would be too apt to draw out his dagger upon every slight occasion.

He was not tall; but of the lowest stature, round faced, olivaster complexion; little eye, round, very black, full of spirit; his hair was black as a raven, but quite white twenty years before he died.

I first saw him at Oxford, 1642, after Edgehill fight, but was then too young to be acquainted with so great a doctor. I remember he came several times to Trinity College to George Bathurst, BD, who had a hen to hatch eggs in his chamber, which they daily opened to discern the progress and way of generation. I had not the honour to be acquainted with him till 1651, being my she cousin Montague's physician and friend. I was at that time bound for Italy (but to my great grief dissuaded by my mother's importunity). He was very communicative, and willing to instruct any that were modest and respectful to him. And in order to my journey, gave me, i.e. dictated to me, what to see, what company to keep, what books to read, how to manage my studies: in short, he bid me go to the fountain head, and read Aristotle, Cicero, Avicenna, and did call the neoterics[29] shit-breeches. He wrote a very bad hand, which (with use) I could pretty well read.

I have heard him say, that after his book of *The Circulation of the Blood* came out, that he fell mightily in his practice, and that 'twas believed by the vulgar that he was crack-brained; and all the physicians were against his opinion, and envied him; many wrote against him, as Dr Primige, Paracisanus, etc. With much ado at last, in about twenty or thirty years' time, it was received in all the universities in the world; and, as Mr Hobbes says in his book *De Corpore*, 'he is the only man, perhaps, that ever lived to see his own doctrine established in his lifetime'.

He understood Greek and Latin poetry well, but was no critic, and he wrote very bad Latin. *The Circuitus Sanguinis*[30] was, as I take it, done into Latin by Sir George Ent, as also his book *de Generatione Animalium*,[31] but a little book in duodecimo against Riolani (I think), wherein he makes out his doctrine clearer, was written by himself, and that, as I take it, at Oxford.

His Majesty King Charles I gave him the wardenship of Merton College in Oxford, as a reward for his service, but the times suffered him not to receive or enjoy any benefit by it.

He was physician, and a great favourite of the Lord High Marshall of England, Thomas Howard, Earl of Arundel and Surrey, with whom he travelled as his physician in his ambassade to the Emperor at Vienna. Mr Wenceslaus Hollar (who was then one of his excellency's gentlemen) told me that, in his voyage, he would still be making of excursions into the woods, making observations of strange trees, and plants, earths, etc., natural things, and sometimes like to be lost, so that my Lord Ambassador would be really angry with him, for there was not only danger of thieves, but also of wild beasts.

He was much and often troubled with the gout, and his way of cure was thus; he would then sit with his legs bare, if it were frost, on the leads of Cockaine House, put them into a pail of water, till he was almost dead with cold, and betake himself to his stove, and so 'twas gone.

He was hot-headed, and his thoughts working would many times keep him from sleeping; he told me that then his way was to rise out of his bed and walk about his chamber in his shirt till he was pretty cool, i.e. till he began to have a horror, and then return to bed, and sleep very comfortably.

I remember he was wont to drink coffee; which he and his brother Eliab did, before coffee houses were in fashion in London.

All his profession would allow him to be an excellent anatomist, but I never heard of any that admired his therapeutic way. I knew several practisers in London that would not have given threepence for one of his bills;[32] and that a man could hardly tell by one of his bills what he did aim at.

He did not care for chemistry, and was wont to speak against them with an undervalue.

It is now fit, and but just, that I should endeavour to undeceive the world in a scandal that I find strongly runs of him, which I have met amongst some learned young men: viz., that he made himself a way to put himself out of his pain, by opium; not but that, had he laboured under great pains, he had been ready enough to have done it; I do not deny that it was not according to his principles upon certain occasions; but the manner of his dying was really, and *bona fide*, thus, viz., the morning of his death about ten o'clock, he went to speak, and found that he had the dead palsy in his tongue; then he saw what was to become of him, he knew there was then no hopes of his recovery, so presently sends for his young nephews to come up to him, to whom he gives one his watch ('twas a minute watch with which he made his experiments); to another, another remembrance, etc. made sign to Sambroke, his apothecary (in Blackfriars), to let him blood in the tongue, which did little or no good; and so he ended his days. His practice was not very great towards his latter end; he declined it, unless to a special friend – e.g. my lady Howland, who had a cancer in her breast, which he did cut off and seared, but at last she died of it.

He rode on horseback with a foot-cloth[33] to visit his patients, his man following on foot, as the fashion then was, which was very decent, now quite discontinued. The judges rode also with their foot-cloths to Westminster Hall, which ended at the death of Sir Robert Hyde, Lord Chief Justice. Anthony, Earl of Shaftesbury, would have revived, but several of the judges being old and ill horsemen did not agree to it.

The scandal aforesaid is from Charles Scarborough's saying that he had, towards his latter end, a preparation of opium and I know not what, which he kept in his study to take, if occasion should serve, to put him out of his pain, and which Sir Charles promised to give him; this I believe to be true; but do not at all believe that he really did give it him. The palsy did give him an easy passport.

Dr Harvey told me, and anyone if he examines himself will find it to be true, that a man could not fancy – truthfully – that he is imperfect in any part that he has teeth, eye, tongue, *spina dorsi*,[34] etc. Nature tends to perfection, and in matters of generation we ought to consult more with our sense and instinct, than our reason, and prudence, fashion of the country, and interest. We see what contemptible products are of the prudent politics,[35] weak, fools and rickety children, scandals to nature and their country. The heralds are fools – *tota errant via*.[36] A blessing goes with a marriage for love upon a strong impulse.

MARY HERBERT, COUNTESS OF PEMBROKE
1555–1621

Mary, countess of Pembroke, was sister to Sir Philip Sidney; married to Henry, the eldest son of William, Earl of Pembroke aforesaid; but this subtle old earl did foresee that his fair and witty daughter-in-law would horn[37] his son and told him so and advised him to keep her in the country and not to let her frequent the court.

She was a beautiful lady and had an excellent wit, and had the best breeding that that age could afford. She had a pretty sharp-oval face. Her hair was of a reddish yellow.

She was very salacious, and she had a contrivance that in the spring of the year the stallions were to be brought before such a part of the house, where she had a *vidette* to look on them. One of her great gallants was crook-backed Cecil, earl of Salisbury.

In her time Wilton House was like a college, there were so many learned and ingenious persons. She was the greatest patroness of wit and learning of any lady in her time. She was a great chemist and spent yearly a great deal in that study. She kept for her laborator in the house Adrian Gilbert (vulgarly called Dr Gilbert), half brother to Sir Walter Raleigh, who was a great chemist in those days. 'Twas he that made the curious wall about Rowlington park, which is the park that adjoins to the house at Wilton. Mr Henry Sanford was the earl's secretary, a good scholar and poet, and who did pen part of the *Arcadia* dedicated to her (as appears by the preface). He has a preface before it with the two letters of his name. She also gave an honourable yearly pension to Dr Thomas Mouffett, who has written a book *De insectis*. Also one Boston, a good chemist, a Salisbury man born, who did undo himself by studying the philosopher's stone, and she would have kept him but he would have all the gold to himself and so died I think in a jail.

At Wilton is a good library which Mr Christopher Wase can give you the best account of any one; which was collected in this learned lady's time. There is a manuscript very elegantly written, viz. all the Psalms of David translated by Sir Philip Sydney, curiously bound in crimson velvet. There is a MS written by Dame Marian of hunting and hawking, in English verse, written in King Henry VIII's time. There is the legder book of Wilton, one page Saxon and the other Latin, which Mr Dugdale perused.

The curious seat of Wilton and the adjacent country is an Arcadian place and a paradise. Sir Philip Sydney was much here, and there was great love between him and his fair sister.

I have heard old gentlemen (old Sir Walter Long of Dracot and old Mr Tyndale) say the first Philip, Earl of Pembroke, inherited not the wit of either the brother or sister.

This countess, after her lord's death, married to Sir Mathew Lister, knight, one of the College of Physicians, London. He was (they say) a learned and a handsome gentleman. She built then a curious house in Bedfordshire called Houghton Lodge near Ampthill. The architects were sent for from Italy. It is built according to the description of Basilius's house in the first book of the *Arcadia* (which is dedicated to her). It is most pleasantly situated and has four vistas, each prospect twenty-five or thirty miles. This was sold to the Earl of Elgin. The house did cost £10,000 the building.

I think she was buried in the vault in the choir at Salisbury, by Henry, Earl of Pembroke, her first husband: but there is no memorial of her, nor of any of the rest, except some pennons and scutcheons.

An epitaph on the lady Mary, countess of Pembroke (in print somewhere), by William Browne, who wrote the *Pastoralls*, whom William, Earl of Pembroke, preferred to be tutor to the first Earl of Carnarvon ([Robert] Dormer), which was worth to him £5,000 or £6,000 i.e. he bought £300 per annum land – from old Jack Markham –

Underneath this sable hearse
Lies the subject of all verse:
Sydney's sister, Pembroke's mother.
Death! er'st thou shalt kill such another
Fair and good and learn'd as shee,
Time will throw a dart at thee.

NICHOLAS HILL
1570–1610

Mr Nicholas Hill: this Nicholas Hill was one of the most learned men of his time: a great mathematician and philosopher and traveller, and a poet. His writings had the usual fate of those not printed in the author's lifetime. He was so eminent for knowledge, that he was the favourite of the great Earl of Oxford, who had him to accompany him in his travels (he was his steward), which were so splendid and sumptuous, that he kept at Florence a greater court than the Great Duke. This earl spent in that travelling, the inheritance of £10,000 or £12,000 per annum.

I fancy that his picture, i.e. head, is at the end of the Long Gallery of Pictures at Wilton, which is the most philosophical aspect that I have seen, very much of Mr T. Hobbes of Malmesbury, but rather *more antique.* 'Tis pity that in noblemen's galleries, the names are not writ on, or behind, the pictures.

He wrote *Philosophia Epicureo-Democritiana, simpliciter proposita, non edocta*: printed at Colen, in an octavo or duodecimo book: Sir John Hoskins has it.

In his travels with his lord (I forget whither Italy or Germany, but I think the former) a poor man begged him to give him *a penny*. 'A penny!' said Mr Hill, 'what dost say to ten pound?' 'Ah! Ten pound!' (said the beggar) 'that would make a man happy.' N. Hill gave him immediately *£10.* And put it down upon account, 'Item, to a beggar £10, to make him happy.'

He printed *Philosophia Epicurea Democritiana*, dedicated *filiolo Laurentio*. There was one Laurence Hill that did belong to the queen's court, that was hanged with Green and Berry about Sir Edmund-Berry Godfrey. According to age, it might be this man, but we cannot be certain.

Mr Thomas Henshaw bought of Nicholas Hill's widow, in Bow Lane, some of his books; among which is a manuscript *De infinitate et aeternitate mundi*. He finds by this writings that he was (or leaning) a Roman Catholic. Mr Henshaw believes he died about 1610: he died an old man. He flourished in Queen Elizabeth's time.

WENCESLAUS HOLLAR
1607–77

Wenceslaus Hollar, a Bohemian, was born at Prague.

His father was a knight of the Empire: which is by letters patent under the imperial seal (as our baronets). I have seen it: the seal is bigger than the broad seal of England: in the middle is the imperial coat; and round about it are the coats of the Princes Elector. His father was a Protestant, and either for keeping a conventicle, or being taken at one, forfeited his estate, and was ruined by the Roman Catholics.

He told me that when he was a schoolboy he took a delight in drawing of maps; which drafts he kept, and they were pretty. He was designed by his father to have become a lawyer, and was put to that profession, when his father's troubles, together with the wars, forced him to leave his country. So that what he did for his delight and recreation only when a boy, proved to be his livelihood when a man.

I think he stayed sometime in low[38] Germany, then he came into England, where he was very kindly entertained by that great patron of painters and draughtsmen (Thomas Howard) Lord High Marshal, Earl of Arundel and Surrey, where he spent his time in drawing and copying rarities, which he did etch (i.e. eat with aqua fortis[39] in copper plates). When the

Lord Marshal went ambassador to the Emperor of Germany to Vienna, he travelled with much grandeur; and among others, Mr Hollar went with him (very well clad) to take views, landscapes, buildings, etc., remarkable to their journey, which we see now at the print shops.

He was very short-sighted and did work so curiously that the curiosity of his work is not to be judged without a magnifying glass. When he took his landscapes, he, then, had a glass to help his sight.

At Arundel House he married with my lady's waiting woman, Mrs Tracy, by whom he has a daughter, that was one of the greatest beauties I have seen; his son by her died in the plague, and ingenious youth, drew delicately.

When the civil wars broke out, the Lord Marshal had leave to go beyond sea. Mr Hollar went into the Low Countries, where he stayed till about 1649.

I remember he told me that when he first came into England (which was a serene time of peace) that the people, both poor and rich, did look cheerfully, but at his return, he found the countenances of the people all changed, melancholy, spiteful, as if bewitched.

I have said before that his father was ruined upon the account of the Protestant religion. Wenceslaus died a Catholic, of which religion, I suppose, he might be ever since he came to Arundel House.

He was a very friendly good-natured man as could be, but shiftless as to the world, and died not rich. He married a second wife, 1665, by whom he has several children. He died on Lady Day, 25th March 1677, and is buried in St Margaret's churchyard at Westminster near the north-west corner of the Tower. Had he lived till 13th July following, he had been just seventy years old.

ROBERT HOOKE
1635–1703

Mr Robert Hooke, curator of the Royal Society at London, was born at Freshwater in the Isle of Wight, 1635; his father was minister there, and of the family of the Hookes of Hooke in Hampshire, in the road from London to Salisbury, a very ancient family and in that place for many (three or more) hundred years.

His father was minister of Freshwater in the Isle of Wight.

He married and had two sons.

John Hoskyns, the painter, being at Freshwater to draw pictures, Mr Hooke observed what he did, and, thought he, 'why cannot I do so too?' So he gets him chalk, and ruddle,[40] and coal, and grinds them and puts them on the trencher, got a pencil, and to work he went, and made a picture: then he copied (as they hung up in the parlour) the pictures there, which he made like. Also, being a boy there, at Freshwater, he made a dial on a round trencher; never having had any instruction. His father was not mathematical at all.

When his father died, his son Robert was left £100, which was sent up to London with him, with an intention to have bound him apprentice to Mr Lilly, the painter, with whom he was a little while upon trial; who liked him very well, but Mr Hooke quickly perceived what was to be done, so, thought he, 'Why cannot I do this by myself and keep my £100?' He also had some instruction from Mr Samuel Cowper (prince of limners of this age).

He went to Mr Busby's, the schoolmaster of Westminster, at whose house he was; and he made very much of him. With him he lodged his £100. There he learned to play twenty lessons on the organ. He there in one week's time made himself master of the first six books of Euclid, to the admiration of Mr Busby, who introduced him. At school here he was very mechanical, and (amongst other things), he invented thirty several ways of flying, which I have heard Sir Richard Knight (who was his school-fellow) say that he seldom saw him in the school.

In 1658 he was sent to Christ Church in Oxford, where he had a chorister's place (in those days when the church music was put down), which was a pretty good maintenance. He was there assistant to Dr Thomas Willis in his chemistry; who afterwards recommended him to the honourable Robert Boyle, esq., to be useful to him in his chemical operations. Mr Hooke then read to him (Robert Boyle) Euclid's *Elements* and made him understand Descartes' philosophy.

In 1662 Mr Robert Boyle recommended Mr Robert Hooke to be curator of the experiments of the Royal Society, wherein he did an admirable good work to the Commonwealth of Learning, in recommending the fittest person in the world to them. In 1664 he was chosen geometry professor at Gresham College.

In 1666 the great conflagration of London happened, and then he was chosen one of the two surveyors of the city of London; by which he has got a great estate. He built Bedlam, the Physicians' College, Montague House, the pillar on Fish Street Hill and theatre there; and he is much made use of in designing buildings.

He is but of middling stature, something crooked, pale faced, and his face but little below, but his head is large; and eye full and popping, and not quick; a grey eye. He has a delicate head of hair, brown, and of an excellent moist curl. He is and ever was very temperate, and moderate in diet, etc.

As he is of prodigious inventive head, so is a person of great virtue and goodness. Now when I have said his inventive faculty is so great, you cannot imagine his memory to be excellent, for they are like two buckets, as one goes up, the other goes down. He is certainly the greatest mechanic this day in the world. His head lies much more to geometry than to arithmetic. He is (1680) a bachelor, and, I believe, will never marry. His elder brother left one fair daughter, which is his heir. *In fine*[41] (which crowns all) he is a person of great suavity and goodness.

'Twas Mr Robert Hooke that invented the pendulum watches, so much more useful than the other watches.

He has invented an engine for the speedy working of division, etc., or for the speedy and immediate finding out the divisor.

WILLIAM LEE
?–1610?

Mr William Lee, MA, was of Oxford, I think, Magdalen Hall.

He was the first inventor of the weaving of stockings by an engine of his contrivance. He was a Sussex man born, or else lived there. He was a poor curate, and, observing how much pains his wife took in knitting a pair of stockings, he bought a stocking and a half, and observed the contrivance in the stitch, which he designed in his loom, which (though some of the appendent instruments of the engine be altered) keeps the same to this day. He went into France, and there died before his loom was made there. So the art was, not long since, in no part of the world but England. Oliver Protector made an act that should be felony to transport this engine. This information I took from a weaver (by this engine) in Pear-poole lane, 1656. Sir John Hoskyns, Mr Stafford Tyndale, and I, went purposely to see it.

SIR JONAS MOORE
1617–79

Sciatica he cured it, by boiling his buttock.

The Duke of York said that 'Mathematicians and physicians had no religion': which being told to Sir Jonas Moore, he presented his duty to the Duke of York and wished with all his heart that his highness 'were a mathematician too': this was since he was supposed to be a Roman Catholic.

He was a clerk under Dr Burghill, Chancellor of Durham. Parson Milbourne, in the bishopric, put him upon the mathematics, and instructed him in it. Then he came to the Middle Temple, London, where he published his *Arithmetic*, and taught it in Stanhope Street. After this, got in with the Lord Gorges, Earl of Bedford, and Sir Thomas Chichiley, for the surveying of the fens – from Captain Sherbourne.

Mr Gascoigne (of the North, I think Yorkshire), a person of good estate, a most learned gentleman, who was killed in the civil wars in the king's cause, a great mathematician, and bred by the Jesuits at Rome, gave him good information in mathematical knowledge.

Sir Jonas Moore was born at Whitelee in Lancashire, towards the bishopric of Durham. He was inclined to mathematics when a boy, and Edmund Wyld, esq., and afterwards Mr Oughtred, more fully informed him; and then he taught gentlemen in London, which was his livelihood.

When the great level of the fens was to be surveyed, Mr Wyld aforesaid who was his scholar and a Member of Parliament was very instrumental in helping him to the employment of surveying it, which was his rise, which I have heard him acknowledge with much gratitude before several persons of quality, since he was a knight, and which evidenced an excellent good nature in him.

When he surveyed the fens, he observed the line that the sea made on the beach, which is not a straight line, by which means he got great credit in keeping out the sea in Norfolk; so he made his banks against the sea of the same line that the sea makes on the beach; and no other could do it, but that the sea would still break in upon it.

He made a model of a citadel for Oliver Cromwell, to bridle the city of London, which Mr Wyld has; and this citadel was to have been the cross building of St Paul's Church.

Upon the restoration of His Majesty he was made Master Surveyor of His Majesty's ordinance and armouries.

In 167[3] he received the honour of a knighthood. He was a good mathematician, and a good fellow.

He died at Godalming, coming from Portsmouth to London, and was buried 2nd September 1679, at the Tower chapel, with sixty pieces of ordinance (equal to the number of years).

He was tall and very fat, thin skin, fair, clear grey eye.

He always intended to have left his library of mathematical books to the Royal Society, of which he was a member; but he happened to die without making a will, whereby the Royal Society have a great loss.

His only son, Jonas, had the honour of knighthood conferred upon him, 9th August 1680, at Windsor; 'His Majesty being pleased to give him this mark of his favour as well in consideration of his own abilities, as of the faithful service of his father deceased' – but young Sir Jonas, when he is old, will never be 'old Sir Jonas', for all the Gazette's eulogy.

I remember Sir Jonas told us that a Jesuit (I think 'twas Grenbergerus, of the Roman College) found out a way of flying, and that he made a youth perform it. Mr Gascoigne taught an Irish boy the way, and he flew over a river in Lancashire (or thereabout), but when he was up in the air, the people gave a shout, whereat the boy being frighted, fell down on the other side of the river, and broke his legs, and when he came to himself, he said that he thought the people had seen some strange apparition, which fancy amazed him. This was in 1635, and he spoke it in the Royal Society, upon the account of the flying at Paris, two years since.

I remember I have heard Sir Jonas say that when he began mathematics, he wonderfully profited by reading Billingsley's *Euclid*, and that 'twas his excellent, clear and plain exposition of the fourth proposition of the first book of the *Elements*, did first open and clear his understanding.

SIR ROBERT MORAY
?–1673

Sir Robert Moray, knight – he was of the ancient family of the Morays in Scotland. The Highlanders (like the Swedes) can make their own clothes; and I have heard Sir Robert say that he could do it.

He spent most of his time in France. After his juvenile education at school and the university he betook himself to military employment in the service of Louis XIII. He was at last Lieutenant-Colonel. He was a great master of the Latin tongue and was very well read. They say he was an excellent soldier.

He was far from the rough humour of the camp breeding, for he was a person the most obliging about the court and the only man that would do a kindness gratis upon an account of friendship. A lackey could not have been more obsequious and diligent. What I do now aver I know to be true upon my own score as well as others. He was a most humble and good man, and as free from covetousness as a Carthusian. He was abstemious and abhorred women. His Majesty was wont to tease at him. 'Twas pity he was a Presbyterian.

He was the chief appuy[42] of his countrymen and their good angel. There had been formerly a great friendship between him and the Duke of Lauderdale, till, about a year or two before his death, he went to the duke on his return from Scotland and told him plainly that he had betrayed his country.

He was one of the first contrivers and institutors of the Royal Society and was our first president, and performed his charge in the chair very well.

He was my most honoured and obliging friend, and I was more obliged to him than to all the courtiers besides. I had a great loss in his death, for, had he lived, he would have got some employment or other for me before this time. He had the king's ear as much as anyone, and was indefatigable in his undertakings. I was often with him. I was with him three hours the morning he died; he seemed to be well enough. I remember he drank at least half a pint of fair water, according to his usual custom.

His lodging where he died was the leaded pavilion in the garden at Whitehall. He died suddenly 4th July about 8 p.m. in 1673. He had but one shilling in his pocket, i.e. *in all.* The king buried him. He lies by Sir William Davenant in Westminster Abbey.

He was a good chemist and assisted His Majesty in his chemical operations.

WILLIAM OUGHTRED
1574–1660

Mr Oughtred – Mr John Sloper tells me that his father was butler of Eton College: he remembers him, a very old man.

A note from my honoured and learned friend Thomas Fludd, esq., who had been High Sheriff of Kent, that he was Mr Oughtred's acquaintance. He told me that Mr Oughtred confessed to him that he was not satisfied how it came about that one might foretell by the stars, but so it was that it fell out true as he did often by his experience find.

Mr William Oughtred, BD, Cambridge, was born at Eton, in Buckinghamshire, near Windsor, in 1574, 5th March, 5 p.m. His father taught to write at Eton, and was a scrivener; and understood common arithmetic, and 'twas no small help and furtherance to his son to be instructed in it when a schoolboy.

His grandfather came from the north for killing a man. The last knight of the family was one Sir Jeffrey Oughtred. I think a Northumberland family.

He was chosen to be one of the king's scholars in Eton College. He went to King's College, in Cambridge.

At the age of twenty-three, he wrote there his *Horologiographia Geometrica*, as appears by the title.[43] He was instituted and inducted into the rectory or parsonage of Albury, in Surrey, worth £100 per annum: he was pastor of this place fifty years.

He married [Mistress] Caryl (an ancient family in those parts), by whom he had nine sons (most lived to be men) and four daughters. None of his sons he could make scholars.

He was a little man, had black hair, and black eyes (with a great deal of spirit). His head was always working. He would draw lines and diagrams on the dust.

His oldest son Benjamin, who lives in the house with my cousin Boothby (who gives him his diet) and now an old man, he bound apprentice to a watchmaker; who did work pretty well, but his sight now fails for that fine work. He told me that his father did use to lie abed till eleven or twelve o'clock, with his doublet on, ever since he can remember. Studied late at night; went not to bed till 11 o'clock; had his tinder box by him; and on the top of his bed-staff, he had his inkhorn fixed. He slept but little. Sometimes he went not to bed in two or three nights, and would not come down to meals till he had found out the *quaesitum*.[44]

He was more famous abroad for his learning, and more esteemed, than at home. Several great mathematicians came over into England on purpose to converse with him. His country neighbours (though they understood not his worth) knew that there must be extraordinary worth in him, that he was visited so by foreigners.

When Mr Seth Ward, MA, and Mr Charles Scarborough, DM, came (as in pilgrimage, to see him and admire him) – they lay at the inn at Shere (the next parish) – Mr Oughtred had against their coming prepared a good dinner, and also he had dressed himself, thus, an old red russet cloth-cassock that had been black in days of yore, girt with an old leather girdle, an old-fashioned russet hat, that had been a beaver,[45] *tempore reginae Elizabethae*.[46] When learned foreigners came and saw how privately he lived, they did admire and bless themselves, that a person of so much worth and learning should not be better provided for.

Seth Ward, MA, a fellow of Sidney College in Cambridge (now bishop of Salisbury) came to him, and lived with him half a year (and he would not take a farthing for his diet), and learned all his mathematics of him. Sir Jonas Moore was with him a good while, and learnt; he was but an ordinary logist before. Sir Charles Scarborough was his scholar; so Dr John Wallis was his scholar; so was Christopher Wren his scholar; so was Mr Smethwyck, FRS. One Mr Austin (a most ingenious man) was his scholar, and studied so much that he became mad, fell a-laughing, and so died, to the great grief of the old gentleman. Mr Stokes, another scholar, fell mad, and dreamed that the good old gentleman came to him and gave him good advice, and so he recovered, and is still well. Mr Thomas Henshawe, FRS, was his scholar (then a young gentleman). But he did not so much like any as those that tugged and took pains to work out questions. He taught all free.

He could not endure to see a scholar write an ill hand; he taught them all presently to mend their hands. Amongst others Mr T.H.[47] who when he came to him wrote a lamentable hand, he taught to write very well. He wrote a very elegant hand, and drew his schemes most neatly, as they had been cut in copper. His father (no doubt) was an ingenious artist at the pen and taught him to write so well.

He was an astrologer, and very lucky in giving his judgements on nativities: he would say, that he did not understand the reason why it should be so, but so it would happen; he did believe that some genius or spirit did help.

He has asserted the rational way of dividing the twelve houses according to the old way, which (the original) Elias Ashmole, esq., has of his own handwriting. Captain George Wharton has inserted it in his Almanack, 1658 or 1659.

The country people did believe that he could conjure and 'tis like enough that he might be well enough contented to have them think so. I have seen some notes of his own handwriting on Cattan's *Geomancy*.

He has told Bishop Ward, and Mr Elias Ashmole (who was his neighbour), that 'on this spot of ground' (or 'leaning against this oak' or 'that ash'), 'the solution of such or such a problem came into my head, as if infused by a divine genius, after I had thought on it without success for a year, two, or three.' Ben Oughtred told me that he had heard his father say to Mr Allen (the famous mathematical instrument maker), in his shop, that he had found out the longitude;[48] *sed vix credo*.[49]

Nicholas Mercator of Holstein went to see him few years before he died. 'Twas about midsummer, and the weather was very hot, and the old gentleman had a good fire, and used Mr Mercator with much humanity, being exceedingly taken with his excellent mathematical wit, and one piece of his courtesy was, to be mighty importunate with him to set on his upper hand next the fire; he being cold with age thought he had been so too.

He was a great lover of chemistry, which he studied before his son Ben can remember, and continued it; and told John Evelyn, of Deptford, esq., FRS, not above a year before he died, that if he were but five years (or three years) younger, he doubted not to find out the philosopher's stone.

He used to talk much of the maiden earth for the philosopher's stone. It was made of the harshest clear water that he could get, which he let stand to putrefy, and evaporated by simmering. Ben tended his furnaces. He has told me that his father would sometimes say that he could make the stone. Quicksilver refined and strained, and gold as it came natural over.

The old gentleman was a great lover of heraldry, and was well known with the heralds at their office, who approved his descent.

He taught a gentleman in half a year to understand Latin, at Mr Duncombe's his parishioner.

His wife was a penurious woman, and would not allow him to burn candle after supper, by which means many a good notion is lost, and many a problem unsolved; so that Mr Thomas Henshawe, when he was there, bought candle, which was great comfort to the old man.

The right honourable Thomas Howard, Earl of Arundel and Surrey, Lord High Marshal of England, was his great patron, and loved him entirely. One time they were like to have been killed together by the fall at Albury of a grotto, which fell down but just as they were come out. My lord had many grottos about his house, cut in the sandy sides of hills, wherein he delighted to sit and discourse.

In the time of the civil wars the Duke of Florence invited him over, and offered him £500 per annum; but he would not accept of it, because of his religion.

Notwithstanding all that has been said of this excellent man, he was in danger to have been sequestered, and [Mr] Onslow that was a great stickler against the royalists and a member of the House of Commons and living not far from him – he translated his *Clavis* into English and dedicated it to him to claw with him, and it did so his business and saved him from sequestration. Now this Onslow was no scholar and hated by the country for bringing his countrymen of Surrey into the trap of slaughter when so many petitioners were killed at Westminster and on the roads in pursuit.

I have heard his neighbour ministers say that he was a pitiful preacher; the reason was because he never studied it, but bent all his thoughts on the mathematics; but when he was in danger of being sequestered for a royalist, he fell to the study of divinity, and preached (they said) admirably well, even in his old age.

He was a good Latinist and Grecian, as appears in a little treatise of his against one Delamaine, a joiner, who was so saucy to write against him (I think about his circles of proportion): upon which occasion I remember I have seen, many years since, twenty or more good verses made, which begin to this purpose:

Thus may some mason or rude carpenter
Put into balance his rule and compasses
'Gainst learned Euclid's pen, etc.

Before he died he burnt a world of papers, and said that the world was not worthy of them; he was so superb. He burned also several printed books, and would not stir, till they were consumed. His son Ben was confident he understood magic.

Mr Oughtred, at the Custom House (his grandson), has some of his papers; I myself have his Pitiscus, embellished with his excellent marginal notes, which I esteem as a great rarity.

I wish I could also have got his Billingsley's *Euclid*, which John Collins says was full of his annotations.

He died 13th June 1660 in the year of his age eighty-eight plus odd days. Ralph Greatorex, his great friend, the mathematical instrument maker, said he conceived he died with joy for the coming-in of the king, which was the 29th May before. 'And are ye sure he is restored?' – 'Then give me a glass of sack to drink His sacred Majesty's health.' His spirits were then quite upon the wing to fly away. The 15th June he was buried in the chancel at Albury, on the north side near the screen. I had much ado to find the very place where the bones of this learned and good man lay (and 'twas but sixteen years after his death). When I first asked his son Ben, he told me that truly the grief for his father's death was so great, that he did not remember the place – now I should have thought it would make him remember it the better – but when he had put on his considering cap (which was never like his father's), he told as aforesaid, with which others did agree.

There is not to this day any manner of memorial for him there, which is a great pity. I have desired Mr John Evelyn, etc., to speak to our patron, the Duke of Norfolk, to bestow a decent inscription of marble on him, which will also perpetuate his grace's fame. I asked Ben concerning the report of his father's dying a Roman Catholic: he told me that 'twas indeed true that when he was sick some priests came from my lord duke's (the Mr Henry Howard, of Norfolk) to him to have discoursed with him, in order to his conversation to their church, but his father was then past understanding. Ben was then by, he told me.

JOHN PELL
1611–85

John Pell, DD, was the son of John Pell, of Southwick in Sussex, in which parish he was born, on St David's day, 1st March 1610 (his young uncle guessed about noon).

His father was a divine but a kind of non-conformist; of the Pells of Lincolnshire, an ancient family; his mother of the Hollands of Kent. His father died when his son John was but five years and six weeks, and left him an excellent library.

He went to school at the free school at Steyning, a borough town in Sussex, at the first founding of the school; with an excellent schoolmaster, John Jeffreys. At thirteen years and a quarter old he went as good a scholar to Cambridge, to

Trinity College, as most Masters of Arts in the university (he understood Latin, Greek and Hebrew), so that he played not much (one must imagine) with his schoolfellows, for, when they had play days, of after-school time, he spent his time in the library aforesaid. He never stood at any election of fellows or scholars of the house at Trinity College.

Of person he was very handsome, and of a very strong and excellent habit of body, melancholic, sanguine, dark brown hair with an excellent moist curl.

Before he went first out of England, he understood these languages (besides his mother tongue), viz., Latin, Greek, Hebrew, Arabic, Italian, French, Spanish, High Dutch and Low Dutch.

In 1632 he married Ithamara Reginalds, second daughter to Mr Henry Reginalds of London. He had by her four sons and four daughters born in this order: S, D, D, S, D, S, D, S.

Dr Pell has said to me that he did believe that he solved some questions *non sine divino auxilio*.[50]

In 1643 he went to Amsterdam, in December; was there professor of mathematics, next after Martinus Hortensius, about two years.

1646, the Prince of Orange called for him to be public professor of philosophy and mathematics at the *Schola Illustris* at Breda, that was founded that year by His Highness.

He returned into England, 1652.

In 1654 Oliver Lord Protector, sent him envoyé to the Protestant cantons of Switzerland; resided chiefly at Zurich. He was sent out with the title of *ablegatus*[51], but afterwards he had order to continue there with the title of 'resident'.

In 1658 he returned into England and so little before the death of Oliver Cromwell that he never saw him since he was Protector.

When he took his leave from Zurich, 23rd June 1658, he made a Latin speech, which I have seen.

In his negotiation he did no disservice to King Charles II, nor to the church, as may appear by his letters which are in the secretary's office.

Richard Cromwell, Protector, did not fully pay him for his business in Piedmont, whereby he was in some want; and so when King Charles II was restored, Dr Sanderson, Bishop of Lincoln, persuaded him to take holy orders. He was not adroit for preaching.

When King Charles II had been home ten months, Mr John Pell first took orders. He was made deacon upon the last of March, 1661, by Bishop Sanderson of Lincoln, by whom he was made priest in June following.

Gilbert Sheldon, Bishop of London, procured for him the parsonage of Fobbing in Essex, 1661, and two years after (1663) gave him the parsonage of Laindon *cum annexa capella de Bartelsdon in codem comitatu,*[52] which benefices are in the infamous and unhealthy (aguish) hundreds of Essex.

Mr Edward Waller on the death of the countess of Warwick:

Curst be already those Essexian plains
Where... Death and Horrour reignes –
etc.

At Fobbing seven curates died within the first ten years; in sixteen years, *six* of those that had been his curates at Laindon are dead; besides those that went away from places; and the death of his wife, servants and grandchildren.

Gilbert Sheldon being made Archbishop of Canterbury, John Pell was made one of his Cambridge chaplains; and complaining one day to his grace at Lambeth of the unhealthiness of his benefice as abovesaid, said my lord 'I do not intend that you shall live there.' 'No,' said Dr Pell, 'but your grace does intend that I shall die there.'

Now by this time (1680), you doubt not but this great, learned man, famous both at home and abroad, has obtained some considerable dignity in the church. You ought not in modesty to guess at less than a deanery – Why, truly, he is staked to this poor preferment still! For though the parishes are large, yet curates, etc., discharged, he clears not about £70 per annum (hardly £80), and lives in an obscure lodging, three stories high, in Jermyn Street, next to the sign of the ship, wanting not only books but his proper MSS, which are many, as by and by will appear. Many of them are at Brereton at my lord Brereton's in Cheshire.

Lord Brereton was sent to Breda to receive the instruction of this worthy person, by his grandfather (George Goring, the Earl of Norwich) in 1647, where he became a good proficient, especially in algebra to which his genius most inclined him and which he used to his dying day, which was 17th March 1680: lies buried in St Martin's Church in-the-fields. I cannot but mention this noble lord but with a great deal of passion, for a more virtuous person (besides his great learning) I never knew. I have had the honour of his acquaintance since his coming from Breda into England. Never was there greater love between master and scholar than between Dr Pell and this scholar of his, whose death 17th March 1680, has deprived this worthy doctor of an ingenious companion and a useful friend.

Dr Pell has often said to me that when he solves a question he strains every nerve about him, and that now in his old age it brings him to a looseness.

Dr J. Pell was the first inventor of that excellent way or method of the marginal working of algebra.

I have heard him say several times that the *Regula falsi* was falsely demonstrated by Mr William Oughtred and that Pitiscus had done it right.

He could not cringe and sneak for preferment though otherwise no man more humble nor more communicative.

He was cast into King's Bench prison for debt 7th September 1680.

In March 1682 he was very kindly invited by Daniel Whistler, MD, to live with him at the Physicians' College in London, where he was very kindly entertained. About the middle of June he fell extreme sick of a cold and removed to a grandchild of his, married to one Mr Hastings in St Margaret's Churchyard, Westminster, near the Tower, who now (1684) lives in Brownlow Street in Drury Lane, where he was like to have been burnt in his bed by a candle. 26th November fell into convulsion fits which had almost killed him.

Gilbert Sheldon, Lord Bishop of London, gave Dr Pell the parsonage of Laindon cum Basildon in the Hundreds of Essex (they call it 'kill-priest' sarcastically) and King Charles II gave him the parsonage of Fobbing, four miles distant. Both are of the value of £200 per annum (or so accounted); but the doctor was a most shiftless man as to worldly affairs, and his tenants and relations couzened him of the profits and kept him so indigent that he wanted necessaries, even paper and ink, and he had not sixpence in his purse when he died, and was buried by the charity of Sir Richard Busby and Dr John Sharp, Rector of St Giles-in-the-fields and Dean of Norwich, who ordered his body to lie in a vault belonging to the rector (the price is £10).

I could not persuade him to make a will; so his books and MSS fell by administratorship to Captain Raven, his son-in-law. His son (John) is a Justice of the Peace in New York, and lives well. He thought to have gone over to him.

This learned person died in St Giles parish aforesaid at the house of Mr Cothorne the reader in Dyot Street on Saturday 12th December 1685, between 4 and 5 p.m. Dr Busby, schoolmaster of Westminster, bought all his books and papers of Captain Raven, among which is the last thing he wrote (which he did at my earnest request), viz. *The Tables*, which are according to his promise in the last line of his printed tables of squares and cubes (if desired) and which Sir Cyrillus Wych (then president of the Royal Society) did license for the press. There only wants a leaf or two for the explanation of the use of them, which his death has prevented. Sir Cyril Wych, only, knows the use of them. I do (imperfectly) remember something of his discourse of them, viz., whereas some questions are capable of several answers, by the help of these tables it might be discovered exactly how many, and no more, solutions, or answers, might be given.

I desired Mr Theodore Haake, his old acquaintance, to make some additions to this short collection of memoirs of him, but he has done nothing.

He died of a broken heart.

Dr Whistler invited Dr Pell to his house, which the doctor likes and accepted of, loving good cheer and good liquor, which the other did also; where eating and drinking too much, was the cause of shortening his days.

Dr Pell had a brother a surgeon and practitioner in physic, who purchased an estate of the natives of New York and when he died he left it to his nephew John Pell, only son of the doctor. It is a great estate eight miles broad.

He had three or four daughters.

SIR WILLIAM PETTY
1632–87

This[53] was done, and a judgement upon it, by Charles Snell, esq., of Alderholt near Fordingbridge in Hampshire – 'Jupiter in Cancer makes him fat at heart.' John Gadbury also says that vomits would be excellent good for him.

Sir William Petty, knight, was the eldest or only son of [Mr] Petty, of Rumsey in Hampshire.

His father was born on the Ash Wednesday, before Mr Hobbes, *scilicet* 1587; and died and was buried at Rumsey 1644, where Sir William intends to set up a monument for him. He was by profession a clothier, and also did dye his own clothes: he left little or no estate to Sir William.

He was born at his father's house aforesaid, on Monday, 26th May 1623, eleven hours 42′56″ afternoon (see Scheme[54]): christened on Trinity Sunday.

Rumsey is a little haven town, but has most kinds of artificers in it. When he was a boy his greatest delight was to be looking on the artificers – e.g. smiths, the watchmaker, carpenters, joiners, etc. – and at twelve years old could have worked at any of these trades. Here he went to school, and learnt by twelve years a competent smattering of Latin, and was entered into the Greek. He has had few sicknesses.

About eight, in April very sick and so continued till towards Michaelmas. About twelve (or thirteen), i.e. before fifteen, he has told me, happened to him the most remarkable 'accident of life' (which he did not tell me), and which was the foundation of all the rest of his greatness and acquiring riches.

He informed me that, about fifteen, in March, he went over into Normandy, to Caen, in a vessel that went hence, with a little stock, and began to merchandise, and had so good success that he maintained himself, and also educated himself; this I guessed was that most remarkable 'accident' that he meant. Here he learnt the French tongue, and perfected himself in the Latin, and had Greek enough to serve his turn.

Here (at Caen) he studied the arts. He was sometime at La Fleche in the college of Jesuits. At eighteen, he was (I have heard him say) a better mathematician than he is now: but when occasion is, he knows how to recur to more mathematical knowledge. At Paris he studied anatomy, and read Vesalius with Mr Thomas Hobbes (see Mr Hobbes' life), who loved his company. Mr Hobbes then wrote his *Optics*; Sir W.P. then had a fine hand in drawing and limning, and drew Mr Hobbes' optical schemes for him, which he was pleased to like. At Paris, one time, it happened that he was driven to a great strait for money, and I have heard him say, that he lived a week on two pennyworth (or three, I have forgot which, but I think the former) of walnuts. He came to Oxford, and entered himself on Brasenose College. Here he taught anatomy to the young scholars. Anatomy was then but little understood by the university, and I remember he kept a body that he brought by water from Reading a good while to read on, some way preserved or pickled.

In 1650 happened that memorable accident and experiment of the reviving Nan Green, a servant maid, who was hanged in the castle of Oxford for murdering her bastard-child. After she had suffered the law, she was cut down, and carried away in order to be anatomised by some young physicians, but Dr William Petty finding life in her, would not venture upon her, only so far as to recover her life. Which being look'd upon as a great wonder, there was a relation of her recovery printed, and at the end several copies of verses made by the young poets of the university were added.

Here he lived and was beloved by all the ingenious scholars, particularly Ralph Bathurst of Trinity College (then doctor of physic); Dr John Wilkins (warden of Wadham College); Seth Ward, DD, astronomy professor; Dr Robert Wood; Thomas Willis, MS, etc.

About these times experimental philosophy first budded here and was first cultivated by the virtuosi in that dark time.

He was about 1650 elected professor of music at Gresham College, by, and by the interest of, his friend Captain John Graunt (who wrote the Observations on the Bills of Mortality), and at that time was worth but £40 in all the world.

Shortly after (that is in 1652 in August, he had the patent for Ireland) he was recommended to the Parliament to be one of the surveyors of Ireland, to which employment Captain John Graunt's interest did also help to give him a lift, and Edmund Wyld, esq., also, then a Member of Parliament, and a great fautor[55] of ingenious and good men, for mere merit's sake (not being formally acquainted with him) did him great service, which perhaps he knows not of.

To be short, he is a person of so great worth and learning, and has such a prodigious working wit, that he is both fit for, and an honour to, the highest preferment.

By this surveying employment he got an estate in Ireland (before the restoration of King Charles II) of £18,000 per annum, the greatest part whereof he was forced afterwards to refund, the former owners being then declared innocents. He has yet there £7,000 or £8,000 per annum and can, from the Mount Mangorton in the county of Kerry, behold fifty thousand acres of his own land.

He has an estate in every province of Ireland.

The kingdom of Ireland he has surveyed, and that with that exactness, that there is no estate there to the value of £60 per annum but he can show, to the value, and those that he employed to the geometrical part were ordinary fellows, some (perhaps) foot soldiers, that circumambulated with their 'box and needles', not knowing what they did, which Sir William knew right well how to make use of.

In 1667, he married on Trinity Sunday the relict[56] of Sir Maurice Fenton, of Ireland, knight, daughter of Sir Hasdras Waller of Ireland, a very beautiful and ingenious lady, brown, with glorious eyes, by whom he has sons and daughters, very lovely children, but all like the mother. He has a natural daughter that much resembles him, no legitimate child so much, that acts at the Duke's playhouse. She is (1680) about twenty-one.

I remember about 1660 there was a great difference between him and one of Oliver's knights. They printed one against the other: this knight was wont to preach at Dublin.

The knight had been a soldier, and challenged Sir William to fight with him. Sir William is extremely short-sighted, and being the challengee it belonged to him to nominate place and weapon. He nominates, for a place, a dark cellar, and the weapon to be a great carpenter's axe. This turned the knight's challenge into ridicule, and so it came to nought.

He can be an excellent droll (if he has a mind to it) and will preach extempore incomparably, either the Presbyterian way, Independent, Capuchin friar or Jesuit.

He received the honour of knighthood. He had his patent for Earl of Kilmore which he stifles during his life to avoid envy, but his son will have the benefit of the precedency. I expected that his son would have broken out a lord or earl: but it seems that he had enemies at the court at Dublin, which out of envy obstructed the passing of his patent.

In 1660 he came into England, and was presently received into good grace with His Majesty, who was mightily pleased with his discourse.

In 1663 he made his double-bottomed vessel (launched about new year's tide), of which he gave a model to the Royal Society made with his own hands, and it is kept in the repository at Gresham College. It did do very good service, but happened to be lost in an extraordinary storm in the Irish Sea. In 1676, 18th March, he was corrupted by the Lord Chancellor Finch, when the patent for the farming of Ireland was sealed, to which Sir William would not seal. Monday 20th March he was affronted by Mr Vernon: Tuesday following Sir William and his lady's brother (Mr Waller) hectored Mr Vernon and carried him.

He went towards Ireland in order to be a member of that parliament, 22nd March 1680 – God send him a prosperous journey.

1680, he went to Rumsey to see his native country, and to erect a monument to his father.

He is a person of an admirable inventive head, and practical parts. He has told me that he has read but little, that is to say, not since he was twenty-five, and is of Mr Hobbes his mind, that he read much, as some men have, he had not known so much as he does, nor should have made such discoveries and improvements.

I remember one St Andrew's day (which is the day of the general meeting of the Royal Society for annual elections), I said, 'methought 'twas not so well that we should pitch upon the patron of Scotland's day, we should rather have taken St George or St Isidore' (a philosopher canonised). 'No,' said Sir

William, 'I would rather have had it on St Thomas' day, for he would not believe till he had seen and put his fingers into the holes,' according to the motto *Nullius in verba*.[57] He has told me that he never got by legacies in his life, but only £10 which was not paid.

He has told me, that whereas some men have accidentally come into the way of preferment, by lying at an inn, and there contracting an acquaintance; on the road; or as some others have done; he never had any such like opportunity, but hewed out his fortune himself.

He is a proper handsome man, measured six foot high, good head of brown hair, moderately turning up: see his picture as doctor of physic. His eyes are a kind of goose-grey, but very short-sighted, and, as to aspect, beautiful, and promise sweetness of nature, and they do not deceive, for he is a marvellous good-natured person, and compassionate.

Eyebrows thick, dark and straight (horizontal). His head is very large. He was in his youth very slender, but since these twenty years and more past he grew very plump, so that now (1680) he is *abdomine tardus*.[58] This last March, 1680, I persuaded him to sit for his picture to Mr Loggan, the graver, whom I forthwith went for myself, and he drew it just before his going into Ireland, and 'tis very like him. But about 1659, he had a picture in miniature drawn by his friend and mine, Mr Samuel Cowper (prince of limners of his age), one of the likest that ever he drew.

I have heard Sir William say more than once that he knew not that he was purblind till his master (a master of a ship) bade him climb up the rope ladder, and give notice when he espied such a steeple (somewhere upon the coast of England or France, I have forgot where), which was a landmark for the avoiding of a shelf; at last the master saw it on the deck, and they fathomed and found they were but [a few] foot water, whereupon (as I remember) his master drubbed him with a cord.

Before he went into Ireland, he solicited, and no doubt he was an admirable good solicitor. I have heard him say that in soliciting (with the same pains) he could dispatch several businesses, nay, better than one alone, for by conversing with several, he should gain the more knowledge, and the greater interest.

Sir William Petty had a boy that whistled incomparably well. He after waited on a lady, a widow, of good fortune. Every night this boy was to whistle the lady asleep. At last she marries him. This is certain true – from himself and Mrs Grant.

FRANCIS POTTER
1594–1678

Mr Francis Potter's father was one of the benefactors to the organ at the cathedral church at Worcester.

Francis Potter, BD, born at Mere, a little market town in Wiltshire, 'upon Trinity Sunday eve 1594, in the evening' – in 1625, 10th December, *horâ decimal inventum est mysterium Bestiae*[59] as he went up the stair to his chamber (which was at his brother's, that is the great room that now is added to the president's lodging).

His father was a minister there, and also of Kilmanton in the county of Somerset about three miles distant, and was also a prebendary of the cathedral church of Worcester. He had three sons, Hannibal, Francis and [another]. His wife's name was Horsey; of the worshipful and ancient family of the Horseys in Clifton in the county of Dorset.

He was taught his grammar learnings by Mr Henry Bright (the famous schoolmaster of those times) of the school at Worcester.

At fifteen, he went to Trinity College in Oxford, where his father (who was an Oxfordshire man born) had been a fellow.

His brother Hannibal was his tutor. Here he was a commoner twenty-seven years, and was senior to all the house but Dr Kettle and his brother.

His genius lay most of all to the mechanics; he had an admirable mechanical invention, but in that dark time wanted encouragement, and when his father died (which was about 1637) he succeeded him in the parsonage of Kilmanton, worth, per annum, about £140. He was from a boy given to drawing and painting. The founder's (Sir Thomas Pope's) picture in Trinity College hall is of his copying.

He had excellent notions for the raising of water; I have heard him say, that he could raise the water at Worcester with less trouble [than presently]; and that he had never seen a water-house engine, but he could invent a better. Kilmanton is on a high hill, and the parsonage well is extraordinary deep.

There is the most ingenious and useful bucket well, that I ever saw. Now, whereas some deep wells have wheels for men or dogs to go within them, here is a wheel of ... foot diameter, with steps (like stairs) to walk on as if you were going up stairs, and an ordinary body's weight draws up a great bucket, which holds a barrel, and the two buckets are contrived so that their ropes always are perpendicular and consequently parallel, and so never interfere with one another. Now, this vast bucket would be too cumbersome to overturn to power out the water; and therefore he contrived a board with lifts about the sides, like a trough, to slide under the bucket, when 'tis drawn up; and at the bottom of the bucket is a plug, the weight of the water jogging upon the sliding trough, the water powers out into the trough, and from thence runs into your pail, or other vessel. 'Tis extremely well worth the seeing. I have taken heretofore a draft of it. I have heard him say that he would have undertaken to have brought up the water from the springs at the bottom of the hill to the town of Shaftesbury, which is on a waterless hill.

In 1625 going into his chamber, the notion of twenty-five, the root of six hundred and sixty-six, for the root of the number of the Beast of the Revelation, came into his head; so he opposed twenty-five to twelve, the root of one hundred and forty-four.

When he took his degree of Bachelor in Divinity, his question was *An Papa sit Anti-Christus*?[60] In his younger years he was very apt to fall into a swoon, and so he did when he was disputing in the Divinity School upon that question. I remember he told me that one time reading Aristotle, *De Natura Animalium*,[61] where he describes how that lionesses, when great with young, and near their time of parturition, do go between two trees that grow near together, and squeeze out their young ones out of their bellies; he had such a strong idea of this, and of the pain that the lioness was in, that he fell into a swoon.

He was of a very tender constitution, and sickly most of his younger years. His manner was, when he was beginning to be sick, to *breathe strongly* a good while together, which he said did emit the noxious vapours.

He was always much contemplative, and had an excellent philosophical head. He was no great read man; he had a competent knowledge in the Latin, Greek and Hebrew tongues, but not a critic. Greek he learnt by Montanus' *Interlineary Testament*, after he was a man, without grammar, and then he read Homer. He understood only common arithmetic, and never went further in geometry than the first six books of Euclid; but he had such an inventive head, that with this foundation he was able to do great matters in the mechanics, and to solve phenomena in natural philosophy. He had but few books, which when he died were sold for fifty-six shillings, and surely no great bargain. He published nothing but his *Interpretation of the Number 666*, in quarto, printed at Oxford, 1642, which has been twice translated into Latin, into French and other languages. He made the final [sun]dial with its furniture, on the north wall of the quadrangle at Trinity College, which he did by Samminitiatus' *Book of Dialling* (it has been gone about 1670, and another is there put). He lived and died a bachelor. He was hospitable, virtuous and temperate; and, as I said before, very contemplative. He looked the most like a monk, or one of the pastors of the old time, that I ever saw one. He was pretty long visaged, and pale clear skin, grey eye. His discourse was admirable, and all new and unvulgar. His house was as undecked as a monk's cell; yet he had there so many ingenious inventions that it was very delightful. He had a pretty contrived garden there, where are the finest box hedges of his planting that ever I saw. The garden is a good large square; in the middle is a good high mount, all fortified (as you may say) and adorned with these hedges, which at the interstices have a high pillar (square cut) of box, that shows very stately and lovely both summer and winter.

On the buttery door in his parlour he drew his father's picture at length, with his book (fore-shortened), and on the spectacles in his hand is the reflection of the gothic south window. I mention this picture the rather, because in process of time it may be mistaken by tradition for his son Francis' picture, author of the book aforesaid.

I never have enjoyed so much pleasure, nor ever so much pleased with such philosophical and hearty entertainment as from him. His book was in the press at Oxford, and he there, when I was admitted of the college, but I had not honour and happiness to be acquainted with him till 1649 (Epiphany), since which time I had a conjunct friendship with him to his death, and corresponded frequently with him. I have all his letters by me, which are very good, and I believe near two hundred, and most of them philosophical.

I have many excellent good notes from him as to mechanics, etc. and I never was with him but I learned and always took notes; but now indeed the Royal Society has outdone most of his things, as having a better apparatus, and more spare money. I have a curious design of his to draw a landscape or perspective (1656), but Sir Christopher Wren has fallen on the same principle, and the engine is better worked.

He was smith and joiner enough to serve his tune, but he did not pretend curiosity for each. He gave me a quadrant in copper, and made me another in silver, of his own projection, which serves for all latitudes. He showed me, 1649, the best way of making an arch was a parabola with a chain; so he took off his girdle from his cassock, and applied it to the wall, thus.

He invented and made with his own hands a pair of beam compasses, which will divide an inch into a hundred or a thousand parts. At one end of the beam is a roundel, which is divided into one hundred equal parts, with a sagitta[62] to turn about it with a handle: this handle turns a screw of a very fine thread, and on the back of the sail and beam is a graduation. With these compasses he made the quadrants aforesaid. He gave me a pair of these compasses, which I showed to the Royal Society at their first institution, which they well liked, and I presented them as a rarity to my honoured friend, Edmund Wyld, esq. There are but two of them in the world.

At the Epiphany, 1649, when I was at his house, he then told me his notion of curing diseases, etc., by transfusion of blood out of one man into another, and that the hint came into his head reflecting on Ovid's story of Medea and Jason, and that this was a matter of ten years before that time. About a year after, he and I went to try the experiment, but 'twas on a hen, and the creature too little and our tools not good: I then sent him a surgeon's lancet. I received a letter from him concerning this subject, which many years since I showed, and was read and entered in the books of the Royal Society, for Dr Lower would have arrogated the invention to himself, and now one R. Griffith, doctor of physic, of Richmond, is publishing a book of the transfusion of blood, and desires to insert Mr Potter's letter.

He was chosen fellow of the Royal Society, and was there admitted and received with much respect.

As he was never a strong man, so in his later times he had his health best, only about four or five years before his death his eyesight was bad, and before he died quite lost. He is buried in the chancel of Kilmanton.

He played at chess as well as most men.

Colonel Bishop, his contemporary at Trinity College, is accounted the best of England. I have heard Mr Potter say that they two have played at Trinity College (I think two days together) and neither got the mastery. He would say that he looked upon the play at chess as very fit to be learnt and practised by young men, because it would make them to have a foresight and be of use to them (by consequence) in their ordering of human affairs.

He has told me that he had oftentimes dreamt that he was at Rome, and being in fright that he should be seized on and brought before the pope, did wake with the fear.

'Twas pity that such a delicate inventive wit should be staked to a private preferment in an obscure corner (where he wanted ingenious conversation), from whence men rarely emerge to a higher preferment, but contract a moss on them like an old pale in an orchard for want of ingenious conversation, which is a great want even to the deepest-thinking men (as Mr Hobbes has often said to me).

The last time I saw this honoured friend of mine, October 1674. I had not seen him in three years before, and his lippitude[63] then was come even to blindness, which did much grieve me to behold. He had let his beard be uncut, which was wont to be but little. I asked him why he did not get some kinswoman or kinsman of his to live with him, and look to him now in his great age? He answered me that he had tried that way, and found it not so well; for they did begrudge what he spent that 'twas too much and went from them, whereas his servants (strangers) were kind to him and took care of him.

In the troublesome times 'twas his happiness never to be sequestered. He was once maliciously informed against to the committee at Wells (a thing very common in those times).

When he came before them, one of them (I have forgotten his name) gave him a pint of wine, and gave him great praise, and bade him go home, and fear nothing.

SIR WALTER RALEIGH
1552–1618

In his youth for several years he was under straits for want of money. I remember that Mr Thomas Child of Worcestershire told me that Sir Walter borrowed a gown of him when he was at Oxford (they were both of the same college), which he never restored, nor money for it.

He went into Ireland, where he served in the wars, and showed much courage and conduct, but he would be perpetually differing with the Lord Deputy; so that at last the hearing was to be at the council table before the queen, which was that he desired; where he told his tale so well and with so good a grace and presence that the queen took especial notice of him and presently preferred him. (So that it must be before this that he served in the French wars.) Queen Elizabeth loved to have all the servants of her court proper men, and (as beforesaid Sir W.R.'s graceful presence was no mean recommendation to him). I think his first preferment at court was captain of Her Majesty's guard. There came a country gentleman (or sufficient yeoman) up to town, who had several sons, but one an extraordinary proper handsome fellow, whom he did hope to have preferred to be a yeoman of the guard. The father (a goodly man himself) comes to Sir Walter Raleigh a stranger to him, and told him that he had brought up a boy that he would desire (having many children) should be one of Her Majesty's guards. Said Sir Walter Raleigh, 'Had you spoken for yourself I should readily have granted your desire, for your person deserves it, but I put in no boys.' Said the father, 'Boy, come in.' The son enters, about eighteen or nineteen, but such a goodly proper young fellow, as Sir Walter Raleigh had not seen the like – he was the tallest of all the guard. Sir Walter Raleigh swears him immediately; and ordered him to carry up the first dish at dinner, where the queen beheld him with admiration, as if a beautiful young giant had stalked in with the service.

He was the first that brought tobacco into England, and into fashion. In our part of North Wilts, e.g. Malmesbury hundred, it came first into fashion by Sir Walter Long.

I have heard my grandfather Lyte say that one pipe was handed from man to man round about the table. They had first silver pipes; the ordinary sort made use of a walnut shell and a straw.

It was sold then for its weight in silver. I have heard some of our old yeomen neighbours say that when they went to Malmesbury or Chippenham market, they culled out their biggest shillings to lay in the scales against tobacco.

Sir W.R., standing in a stand at Sir Robert Poyntz's park at Acton, took a pipe of tobacco, which made the ladies quit it till he had done.

Within these thirty-five years 'twas scandalous for a divine to take tobacco.

Now, the customs of it are the greatest His Majesty has. Rider's Almanac – 'Since tobacco brought into England by Sir Walter Raleigh, ninety-nine years, the custom whereof is now the greatest of all others and amounts to yearly...' Mr Michael Weekes of the Royal Society assures me, out of the custom-house books that the custom of tobacco over all England is £400,000 per annum.

He was a tall, handsome and bold man: but his naeve[64] was that he was damnable proud. Old Sir Robert Harley of Brampton Brian castle, who knew him, would say 'twas a great question who was the proudest, Sir Walter, or Sir

Thomas Overbury, but the difference that was, was judged on Sir Thomas' side.

His beard turned up naturally. I have heard my grandmother say that when she was young, there were wont to talk of this rebus, viz.

The enemy to the stomach, and the word of disgrace,
Is the name of the gentleman with a bold face.

Old Sir Thomas Malett, one of the justices of the King's Bench *tempore Caroli I et II*,[65] knew Sir Walter; and I have heard him say that, notwithstanding his so great mastership in style and his conversation with the learnedest and politest persons, yet he spoke broad Devonshire to his dying day.

His voice was small, as likewise were my schoolfellows', his grand-nephews.

Sir Walter Raleigh was a great chemist; and amongst some MSS receipts I have seen some secrets from him. He studied most in his sea voyages, where he carried always a trunk of books along with him, and had nothing to divert him.

He made an excellent cordial, good in fevers, etc. Mr Robert Boyle has the recipe, and makes it and does great cures by it.

A person so much immersed in action all along and in fabrication of his own fortunes (till his confinement in the Tower) could have but little time to study, but what he could spare in the morning. He was no slug; without doubt, had a wonderful walking spirit, and great judgement to guide it.

Durham House was a noble palace; after he came to his greatness he lived there, or in some apartment of it. I well remember his study, which was a little turret that looked into and over the Thames, and had the prospect which is pleasant perhaps as any in the world, and which not only refreshes the eyesight but cheers the spirits, and (to speak my mind) I believe enlarges an ingenious man's thoughts.

Sherborne Castle, park, manor, etc., did belong (and still ought to belong) to the church of Salisbury. Then Sir Walter Raleigh begged it as a boon from Queen Elizabeth: where he built a delicate lodge in the park, of brick, not big, but very convenient for the bigness, a place to retire from the court in the summer time, and to contemplate, etc. Upon his attainder, 'twas begged by the favourite Carr, Earl of Somerset, who forfeited it (I think) about the poisoning of Sir Thomas Overbury. Then John, Earl of Bristowe, had it given him for his good service in the ambassade in Spain, and added two wings to Sir Walter Raleigh's lodge. In short and indeed 'tis a most sweet and pleasant place and site as any in the west, perhaps none like it.

In his youth his companions were boisterous blades, but generally that had wit; except otherwise upon design to get them engaged for him – e.g. Sir Charles Snell, of Kington Saint Michael in North Wilts, my good neighbour, an honest young gentleman but kept a perpetual sot, he engaged him to build a ship (the Angel Gabriel) for the design of Guyana, which cost him the manor of Yatton-Keynell, the farm at Easton Piers, Thornhill and the church lease of Bishops Cannings; which ship, upon Sir Walter Raleigh's attainder, was forfeited.

In his youthful time was one Charles Chester, that often kept company with his acquaintance; he was a bold impertinent fellow, and they could never be at quiet for him; a perpetual talker, and made a noise like a drum in a room. So one time at a tavern Sir W.R. beats him and seals up his mouth (i.e. his upper and nether beard) with hard wax. From him Ben Jonson takes his Carlo Buffono (i.e. 'jester') in *Every Man out of his Humour*.

He was a second to the Earl of Oxford in a duel. Was acquainted and accepted with all the heroes of our nation in his time.

Sir Walter Long, of Draycot (grandfather to this old Sir James Long), married a daughter of Sir John Thynne, by which means, and their consimility of disposition, there was a very conjunct friendship between the two brothers (Sir Carew and Sir Walter) and him; and old John Long, who then waited on Sir W. Long, being one time in the privy garden with his master, saw the Earl of Nottingham wipe the dust from Sir Walter R.'s shoes with his cloak, in compliment.

In the great parlour of Downton, at Mr Raleigh's, is a good piece (an original) of Sir W. in a white satin doublet, all embroidered with rich pearls about his neck, and the old servants have told me that the pearls were near as big as the painted ones.

He had a most remarkable aspect, and exceeding high forehead, long-faced, and sour eye-lidded, a kind of pig eye.

At an obscure tavern, in Drury Lane (a bailiff's), is a good picture of this worthy, and also of others of his time; taken upon some execution (I suppose) formerly.

I have heard old Major Cosh say that Sir W. Raleigh did not care to go on the Thames in a wherry boat: he would rather go round about over London Bridge.

My old friend James Harrington, esq., was well acquainted with Sir Benjamin Ruddyer, who was an acquaintance of Sir Walter Raleigh's. He told Mr J.H. that Sir Walter Raleigh being invited to dinner to some great person where his son was to go with him, he said to his son, 'Thou art expected today at dinner to go along with me, but thou art such a quarrelsome, affronting [man], that I am ashamed to have such a bear in my company.' Mr Walter humbled himself to his father, and promised he would behave himself mighty mannerly. So away they went (and Sir Benjamin, I think, with them). He sat next to his father and was very demure at least half dinner time. Then said he, 'I, this morning, not having the fear of God before my eyes but by the instigation of the devil, went …' Sir Walter being strangely surprised and put out of his countenance at so great a table, gives his son a damned blow over the face. His son, as rude as he was, would not strike his father, but strikes over the face the gentleman that sat next to him and said, 'Box about: 'twill come to my father anon.' 'Tis now a common-used proverb.

He loved one of the maids of honour. She proved with child and I doubt not but this hero took care of them both, as also that the product was more than an ordinary mortal.

'Twas Sir Walter Raleigh's epigram on Robert Cecil, Earl of Salisbury, who died in a ditch three or four miles west from Marlborough, returning from Bath to London, which was printed in an octavo book about 1656 –

Here lies Robert, our shepherd whilere,
Who once in a quarter our fleeces did sheer:
For his oblation to Pan his manner was thus,
He first gave a trifle, then offered up us.
… … … … …
In spite of the tarbox he died of the shabbo.[66]

This I had from old Sir Thomas Malett, one of the judges of the King's Bench, who knew Sir Walter Raleigh, and did remember these passages.

I have now forgot whether Sir Walter was not for the putting of Mary, Queen of Scots, to death; I think, yea.

But, besides that, at a consultation at Whitehall, after Queen Elizabeth's death, how matters were to be ordered and what ought to be done, Sir Walter Raleigh declared his opinion, 'twas the wisest way for them to keep the government in their own hands, and set up a commonwealth, and not be subject to a needy beggarly nation. It seems there were some of this cabal who kept not this so secret but that it came to King James's ear; who where the English noblesse met and received him, being told upon their presentment to His Majesty their names, when Sir Walter Raleigh's name was told 'Raleigh' said the king, 'On my soul, man, I have heard *rawly* of thee.'

He was such a person (every way) that (as King Charles I says of the lord Strafford) a prince would rather be afraid than ashamed of. He had that awfulness and ascendancy in his aspect over other mortals. It was a most stately sight, the glory of that reception of His Majesty, where the nobility and gentry were in exceeding rich equipage, having enjoyed a long peace under the most excellent of queens; and the company was so exceedingly numerous that their obedience carried a secret dread with it.

King James did not inwardly like it, and with an inward envy said that, though so and so (as before), he doubted not but he should have been able on his own strength (should the English have kept him out) to have dealt with them, and get his right. Said Sir Walter Raleigh to him, 'Would to God that had been put down to trial.' 'Why do you wish that?' said the king.

'Because,' said Sir Walter, 'that then you would have known your friends from your foes.' But that reason of Sir Walter was never forgotten nor forgiven.

Old Major Stansby, a most intimate friend and neighbour and coetanean[67] of the late Earl of Southampton (Lord Treasurer), told me from his friend, the earl, that as to the plot and business about the lord Cobham, etc., he being then governor of Jersey, would not fully do things unless they would go to his island and there advise and resolve about it; and that really and indeed Sir Walter's purpose was when he had them there, to have betrayed them and the plot, and to have them delivered up to the king and made his peace.

Mr Edmund Wyld knew him and says he was a learned and sober gentleman and good mathematician, but if you happened to speak of Guyana he would be strangely passionate and say 'twas 'the blessedest country under the sun', etc., reflecting on the spoiling that brave design.

When he was attached by the officer about the business which cost him his head, he was carried in a wherry, I think only with two men. King James was wont to say that he was a coward to be so taken and conveyed, for else he might easily have made his escape from so slight a guard.

He was prisoner in the Tower. There (besides his compiling his *History of the World*) studied chemistry. The Earl of Northumberland was prisoner at the same time, who was the patron to Mr Hariot and Mr Warner, two of the best mathematicians then in the world, as also Mr Hues (who wrote *De Globis*). Sergeant Hoskins (the poet) was a prisoner there too.

I heard my cousin Whitney say that he saw him in the Tower. He had a velvet cap laced, and a rich gown, and trunk hose.

He was scandalised with[68] atheism; but he was a bold man, and would venture at discourse which was unpleasant to the churchmen. I remember the first lord Scudamore said 'twas basely said of Sir W.R. to talk of 'the anagram of Dog'. In his speech on the scaffold, I heard my cousin Whitney say (and I think 'tis printed) that he spoke not one word of Christ, but of the great and incomprehensible God, with much zeal and adoration, so that he concluded he was an a-Christ,[69] not an atheist.

He took a pipe of tobacco a little before he went to the scaffold, which some formal persons were scandalised at, but I think 'twas well and properly done to settle his spirits.

I remember I heard old Father Symonds (of the Jesuits) say, that a father was at his execution, and that to his knowledge he died with a lie in his mouth: I have now forgot what 'twas. The time of his execution was contrived to be on my Lord Mayor's day (viz. the day after St Simon and Jude) 1618, that the pageants and fine shows might draw away the people from beholding the tragedy of one of the gallantest worthies that ever England bred. Buried privately under the high altar at St Margaret's Church, in Westminster; in which grave (or near) lies James Harrington, esq., author of *Oceana*.

Mr Elias Ashmole told me that his son Carew Raleigh told him he had his father's skull; that some years since, upon digging up the grave, his skull and neckbone being viewed, they found the bone of his neck lapped over so, that he could not have been hanged.

Baker's *Chronicle* – 'A scaffold was erected in the Old Palace Yard, upon which, after fourteen years' reprieve, his head was cut off. At which time such abundance of blood issued from his veins that showed he had stock of nature enough left to have continued him many years in life though now above three score years old, if it had not been taken away by the hand of violence. And this was the end of the great Sir Walter Raleigh, great sometimes in the favour of Queen Elizabeth, and (next to Sir Francis Drake) the great scourge and hate of the Spaniard; who had many things to be commended in his life, but none more than his constancy at his death, which he took with so undaunted a resolution that one might perceive he had a certain expectation of a better life after it, so far he was from holding those atheistic opinions, an aspersion whereof some had cast upon him.'

He was buried as soon as you are removed from the top of the steps towards the altar, not under the altar.

On Sir Walter Raleigh
Here lieth, hidden in this pitt,
The wonder of the world for witt.
It to small purpose did him serve;
His witt could not his life preserve.
Hee living was belov'd of none,
Yet in his death all did him moane.
Heaven hath his soule, the world his fame,
The grave his corps, Stukley his shame.

This I found among the papers of my honoured friend and neighbour Thomas Tyndale, esq.

At the end of the *History of the World*, he laments the death of the most noble and most hopeful prince Henry, whose great favourite he was, and who, had he survived his father, would quickly have enlarged him, with rewards of honour. So upon the prince's death ends his first part of his *History of the World*, with a gallant eulogy of him, and concludes, *Versa est in luctum cithara mea; et cantus meus in vocem flentium.*[70]

His book sold very slowly at first, and the bookseller complained of it, and told him that he should be a loser by it, which put Sir W. into a passion; and said that since the world did not understand it, they should not have his second part, which he took and threw into the fire, and burnt before his face.

Mr Elias Ashmole says that Degore Whear in his *Praelectiones Hyemales* gives him an admirable encomium, and prefers him before all other historians.

He was sometimes a poet, not often. Before Spenser's *Faerie Queene* is a good copy of verses, which begins thus, 'Methinks I see the grave where Laura lay', at the bottom W.R.: which, thirty-six years since, I was told were his.

A copy of Sir W. Raleigh's letter, sent to Mr Duke, in Devon, written with his own hand.

Mr Duke,

I wrote to Mr Prideaux to move you for the purchase of Hayes, a farm sometime in my father's possession. I will most willingly give whatsoever in your conscience you shall deem it worth, and if at any time you shall have occasion to use me, you shall find me a thankful friend to you and yours. I am resolved, if I cannot but entreat you, to build at Colliton; but for the natural disposition I have to that place, being born in that house, I had rather seat myself there than anywhere else; I take my leave, ready to countervail all your courtesies to the utter of my power.

Your very willing friend, in all I shall be able,
Walter Raleigh

Even such is tyme, which takes in trust
Our youth, our joyes, and all we have,
And payes us but with age and dust.
Within the darke and silent grave,
When we have wandered all our wayes,
Shutts up the story of our dayes.
But from which grave and earth and dust
The Lord will rayse me up I trust.

These lines Sir Walter Raleigh wrote in his Bible, the night before he was beheaded, and desired his relations with these words, viz. 'Beg my dead body, which living is denied you; and bury it either in Sherborne or Exeter church.'

ROBERT RECORD
1510?–58

He was the first that wrote a good arithmetical treatise in English, which has been printed a great many times, viz. his 'Arithmetick, containing the ground of arts in which is taught the general part rules and operations of the same in whole numbers and fractions after a more easie and exact method then ever heretofore.'

It was dedicated 'to the most mighty prince Edward the 6th by the grace of God king of England, Scotland, France and Ireland, etc.' In the end of which epistle:

> How some of these statutes may be applied to use as well in our time as in any other time I have particularly declared in this book and some other I have omitted for just considerations till I may offer them first unto your majestie to weigh them as to your highness shall seem good. For many things in them are not to be published without your highness knowledge and approbation, namely because in them is declared all the rates of all oyles, for all standards from an ounce upwards, with other mysteries of mint-matters, and also most part of the varieties of coins that have been current in this realm by the space of 600 years last past, and many of them were current in the time that the Romans ruled here. All which with the ancient description of England and Ireland, and my simple censure of the same, I have almost completed to be exhibited to your highness.

> To the reader: – It shall induce me to set forth those further instructions concerning geometrie and cosmography which I have already promised and am sure has not hitherto in our English tongue been published.

The Whetstone of Witt, which is the second part of *Arithmetick*, containing the extraction of roots, the cossick practice, with the rule of equation and the works of surd numbers. Quarto; dedicated 'to the right worshipful the governors, consuls, and the rest of the company of venturers into Muscovia'. Here he speaks:

> For your commodities I will shortly set forth such a book of navigation as I dare say shall partly satisfy and content not only your expectation but also the desire of a great number besides. Wherein I will not forget specially to touch both the old attempt for the northerly navigations and the late good adventure with the fortunate success in discovering that voyage which no man before you durst attempt since the time of king Alfred's reign, I mean by the space of 700 years, never any before that time had passed that voyage except one other that dwelt in Halgolande who reported that journey to the noble king Alurede, as it does yet remain in ancient records of the old Saxon tongue. In that book also I will show certain means how without great difficulty you may sail to the North-East Indies and so to Camul Chinchital and Balor which are countries of great commodities; as for Chatai lies so far within the land toward the South Indian seas that the journey is not to be attempted until you are better acquainted with those countries that you must first arrive at. – At London the xii day of November 1557.

In the last leaf of this book he is frighted by the hasty knocking of a messenger at the door and says:

> Then there is no remedy but that I must neglect all studies and teaching to withstand these dangers. My fortune is not so good to have quiet time to teach.

The Castle of Knowledge, printed at London, 1596, quarto, and is dedicated 'to the most mightie and most puissant princesse, Marie, by the grace of God, Queen of England, Spain, both Sicilies, France, Jerusalem, and Ireland, Defender of the Faith, Archduchesse of Austria, Duchesse of Milaine, Burgundy, and Brabant, Countesse of Habsburg, Flanders, Tyroll, etc.'

He was the first that ever wrote of astronomy in the English tongue.

In an admonition for orderly studying of the author's works before this book there is an intimation in verse that he wrote these five books, *scilicet*, (1) *The Ground of Arts*, (2) *The Pathway to Knowledge*, (3) *The Gate of Knowledge*, (4) *The Castle of Knowledge*, (5) *The Treasury of Knowledge*.

All that I have seen of his are written in dialogues between the master and scholar.

ROBSON

Mr [Fabian] Philips also tells me that Robson was the first that brought into England the art of making Venice glasses, but Sir Edward Zouch (a courtier and drolling favourite of King James) oppressed this poor man Robson, and forced it from him, by these four verses to King James, which made his majesty laugh so that he was ready to beshit his briggs. The verses are these:

Severn, Humber, Trent and Thames
And thy great Ocean and her streames
Must putt downe Robson and his fires
Or downe goes Zouche and his desires.

The king granted this ingenious manufacture to Zouch, being tickled as aforesaid with these rhymes, and so poor Robson was oppressed and totally undone, and came to that low degree of poverty that Mr Philips told me that he swept the yard at Whitehall and that he himself saw him do it.

Sir Robert Mansell had the glass-work afterwards, and employed Mr James Howell (author of *The Vocall Forest*) at Venice as a factor to furnish him with materials for his work.

WALTER RUMSEY
1584–1660

Walter Rumsey, of Lanover, in Monmouth, esq. (born there), was of Gloucester Hall in Oxford; afterwards of the society of Gray's Inn, where he was a bencher.

He was one of the judges in south Wales, viz. Carmarthen, Pembrokeshire, and Cardigan circuit. He was so excellent a lawyer, that he was called *The Picklock of the Law*.

He was an ingenious man, and had a philosophical head; he was most curious for grafting, inoculating, and planting, and ponds. If he had any old dead plum-tree, or apple-tree, he let them stand, and planted vines at the bottom, and let them climb up, and they would bear very well.

He was one of my counsel in my lawsuits in Breconshire about the entail. He had a kindness for me and invited me to his house, and told me a great many fine things, both natural and antiquarian.

He was very facetious, and a good musician, played on the organ and lute. He could compose.

He was much troubled with phlegm, and being so one winter at the court of Ludlow (where he was one of the counsels), sitting by the fire, spitting and spawling, he took a fine tender sprig, and tied a rag at the end, and conceited he might put it down his throat, and fetch up the phlegm, and he did so. Afterwards he made this instrument of whalebone. I have often seen him use it. I could never make it go down my throat, but for those that can 'tis a most incomparable engine. If troubled with the wind it cures you *immediately*. It makes you vomit without any pain, and besides, the vomits of apothecaries have *aliquid veneni*[71] in them. He wrote a little octavo book, of this way of medicine, called *Organon Salutis.* I had a young fellow (Marc Collins), that was my servant, that used it incomparably, more easily than the judge; he made of them. In Wilts, among my things are, some of his making still. The judge said he never saw anyone use it so dextrously in his life. It is no pain, when down your throat; he would touch the bottom of his stomach with it.

THOMAS STREET
1621–89

Mr Thomas Street, astronomer, was born in Ireland, his widow thinks, at Castle Lyons, 5th March 1621.

In 1661 he printed that excellent piece of *Astronomica Carolina*, which he dedicated to King Charles II, and also presented it well bound to Prince Rupert and the Duke of Monmouth, but never had a farthing of any of them.

He had the true motion of the moon by which he could do it (he has finished the tables of the moon and also of Mercury, which was never made perfect before) – but two of his familiar acquaintance tell me that he did not commit this discovery to paper: so it is dead with him. He made attempts to be introduced to King Charles II and also to King James II, but courtiers would not do it without a good gratuity.

He was of a rough and choleric humour. Discoursing with Prince Rupert, His Highness affirmed something that was not according to art; said Mr Street, 'whoever affirms that is no mathematician'. So they would point at him afterwards at court and say, 'There's the man that huffed Prince Rupert.'

He has left his widow (who lives in Warwick) an absolute piece of trigonometry, plain and spherical, in MS, more perfect than ever was yet done, and more clear and demonstrated.

He died in Cannon Row (vulgarly Channel Row) at Westminster, 17th August 1689, and is buried in the churchyard of the new chapel there towards the east window of the chancel, that is, within twenty or thirty foot of the wall.

He made this following epitaph himself:

Here lies the earth of one that thought some good,
Although too few him rightly understood:
Above the starres his heightned mind did fly,
His hapier spirit into Eternity.

His acquaintance talk of clubbing towards an inscription. No man living has deserved so well of astronomy.

SETH WARD
1617–89

Seth Ward, Lord Bishop of Salisbury, was born at Buntingford, a small market town in Hertfordshire, in 1618 (when the great blazing star appeared). His father was an attorney there, and of a very honest repute.

At sixteen years old he went to Sidney Sussex College in Cambridge; he was servitor to Dr Samuel Ward (master of the college and professor of divinity), who, being much taken with his ingenuity and industry, as also with his suavity of nature, quickly made him scholar of the house, and after, fellow.

Though he was of his name, he was not at all akin to him (which most men imagined because of the great kindness to him); but the consimility of their dispositions was a greater tie of friendship than that of blood, which signifies but little, as to that point.

His father taught him common arithmetic, and his genius lay much to the mathematics, which being natural to him, he quickly and easily attained.

Sir Charles Scarborough, MD (then an ingenious young student, and fellow of Caius College in Cambridge), was his great acquaintance; both students in mathematics; which the better to perfect, they went to Mr William Oughtred, at Albury in Surrey, to be informed by him in his *Clavis Mathematica*, which was then a book of enigmata.[72] Mr Oughtred treated them with exceeding humanity, being pleased at his heart when an ingenious young man came to him that would ply his algebra hard. When they returned to Cambridge, they read the *Clavis Mathematica* to their pupils, which was the first time that that book was ever read in a university. Mr Laurence Rooke, a good mathematician and algebrist (and I think had also been Mr Oughtred's disciple), was his great acquaintance. Mr Rooke (I remember) did read (and that admirably well) on the sixth chapter of the *Clavis Mathematica* in Gresham College.

In 1644, at the breaking out of the civil wars, he was a prisoner, together with Dr Samuel Collins, Sir Thomas Hatton, etc. for the king's cause, in St John's College in Cambridge, and was put out of his fellowship at Sidney Sussex College. Being got out of prison, he was very civilly and kindly received by his friend and neighbour, Ralph Freeman, of Apsten, esq., a virtuous and hospitable gentleman. In 1648 the visitation of the Parliament was Oxford, and turned out a great many professors and fellows. The astronomy reader (Dr John Greaves) being sure to be ejected, Seth Ward, MA (living then with my lord Wenman, in Oxfordshire, and Greaves was unwilling to be turned out of his place, but desired to resign it rather to some worthy person, whereupon Dr Charles Scarborough and William Holder, DD, recommended to Greaves, their common friend, Mr Seth Ward) was invited to succeed him, and came from Mr Freeman's to Oxford, had the astronomy professor's place, and lived at Wadham College, where he conversed with the warden, Dr John Wilkins.

In 1659 William Hawes, then President of Trinity College in Oxford, having broken in his lungs a vein (which was not curable), Mr Ward being very well acquainted and beloved in that college; by the consent of all the fellows, William Hawes resigned up his presidentship to him, and died some few days after. In 1660, upon the restoration of King Charles II, Dr Hannibal Potter (the president sequestered by the parliamentary visitors) re-enjoyed the presidentship again. Dr Seth Ward, now Bishop of Salisbury, when he was President of Trinity College, Oxford, did draw his geometrical schemes with black, red, yellow, green and blue ink to avoid the perplexity of A, B, C, etc.

In 1661, the Dean of Exeter died, and then it was his right to step in next to the deanery.

In 1663, the Bishop of Exeter died: Dr Ward, the dean, was in Devonshire at that time, at, I think, Tavistock, at a visitation, where were a great number of the gentry of the country. Dean Ward was very well known to the gentry, and his learning, prudence and comity had won them all to be his friends. The news of the death of the bishop being brought to them, who were all very merry and rejoicing with good entertainment, with great alacrity the gentlemen cried all, *uno ore*,[73] 'We will have Mr Dean to be our bishop.' This was at that critical time when the House of Commons were the king's darlings. The dean told them that for his part he had no interest or acquaintance at Court; but intimated to them how much the king esteemed the Members of Parliament (and a great many parliament men were then there), and that His Majesty would deny them nothing. 'If 'tis so, gentlemen' (said Mr Dean), 'that you will needs have me to be your bishop, if some of you will make your addresses to His Majesty, 'twill be done.' With that they drank the other glass, a health to the king, and another to their wished-for bishop; had their horses presently made ready, put foot in stirrup, and away they rode merrily to London; went to the king, and he immediately granted them their request. This is the first time that ever a bishop was made by the House of Commons. Now, though envy cannot deny, that this worthy person, was very well worthy any preferment could be conferred on him, yet the old bishops (e.g. Humphrey Henchman, Bishop of London; John Cosins, Bishop of Durham, etc.) were exceedingly disgruntled at it, to see a brisk young bishop that could see through all their formal gravity, but forty years old, not come in at the right door but leap over the pale. It went to their very hearts. Well, Bishop of Exeter he was, to the great joy of all the diocese.

Being bishop he had then free access to His Majesty, who is a lover of ingenuity and a discerner of ingenious men, and quickly took a liking to him.

In 1667, Alexander Hyde, the Bishop of Salisbury, died and then he was made Bishop of Salisbury, in the month of September.

He is (without all manner of flattery) so prudent, learned, and good a man, that he honours his preferment as much as the preferment does him; and is such a one that cannot be advanced too high. My lord Lucius Falkland was wont to say that he never knew anyone that a pair of lawn sleeves[74] had not altered from himself, but only Bishop Juxon; had he known this excellent prelate, he would have said he had known one more. As he is the pattern of humility and courtesy, so he knows when to be severe and austere; and he is not one to be trampled or worked upon. He is a bachelor, and of a most magnificent and munificent mind.

He has been a benefactor to the Royal Society (of which he was one of the first members and institutors). He also gave a noble pendulum clock to the Royal Society (which goes a week), to perpetuate the memory of his dear and learned friend, Mr Laurence Rooke.

He gave in [many] pounds towards the making of the river at Salisbury navigable to Christ Church. In 1679 he gave to Sidney Sussex College £1,000.

He has perused all the records of the church of Salisbury, which, with long lying, had been conglutinated together; read them all over, and taken abridgements of them, which has not been done by any of his predecessors I believe for some hundreds of years.

He had an admirable habit of body (athletic, which was a fault), a handsome man, pleasant and sanguine; he did not desire to have his wisdom be judged by the gravity of his beard, but his prudence and ratiocination. This, methinks, is strange to consider in him, that being a great student (and that of mathematics and difficult knotty points, which does use to make men unfit for business), he is so clear and ready, as no solicitor is more adroit for looking after affairs.

The black malice of the Dean of Salisbury – he printed sarcastic pamphlets against him – was the cause of his disturbed spirit, whereby at length he quite lost his memory. For about a month before he died he took very little sustenance, but lived on the stock and died a skeleton. He deceased at his house at Knightsbridge near London, on Sunday morning, 6th January 1689: the gazettes and newsletters were severally mistaken as to the day of his death.

I searched all Seth, Bishop of Salisbury's, papers that were at his house at Knightsbridge where he died. I have taken care with his nephew and heir to look over his papers in his study at Salisbury. He tells me the custom is, when the Bishop of Salisbury dies, that 'the dean and chapter lock up his study and put a seal on it'. It was not opened lately, but when it is he will give me an account.

Seth Ward, Lord Bishop of Salisbury, studied the common law, and I find this paper, which is his own handwriting, amongst his scattered papers which I rescued from being used by the cook since his death, which was destined with other good papers and letters to be put under pies.

He wrote a reply to Bullialdus, which might be about the bigness of his *Astronomia Geometrica*, which he lent to somebody and is lost. In the bishop's study are several letters between Bullialdus and him, and between Helvelius and him.

At Buntingford, Hertfordshire: 'This hospital was erected and endowed by Seth Ward, DD, Lord Bishop of Salisbury and chancellor of the most noble order of the garter, who was born in this town within the parish of Aspden and educated in the free school of Buntingford.'

Whereas I put down, from his own mouth, viz., that he said, occasionally, that 'he was born when the great comet appeared' (that, I am sure, was in 1618); but his nephew, Seth Ward, treasurer of the last church of Salisbury and his executor, told me that the last summer he searched in the register at Buntingford where he was born, and finds thus: Seth Ward christened 5th April 1617.

JOHN WILKINS
1614–72

John Wilkins, Lord Bishop of Chester; his father was a goldsmith in Oxford. Mr Francis Potter knew him very well, and was wont to say that he was a very ingenious man, and had a very mechanical head. He was much for trying of experiments, and his head ran much upon the perpetual motion. He married a daughter of Mr John Dod (who wrote on the commandments), at whose house, at Fawsley, near Daventry, Northamptonshire, she lay in with her son John, of whom we are now to speak.

He had a brother (Timothy), squire-beadle of divinity in Oxford, and a uterine brother, Walter Pope, MD.

He had his grammar learning in Oxford (I think from Mr Sylvester). He was admitted of Magdalen Hall in Oxford

(1627). His tutor there was the learned Mr John Tombs (Coryphaeus of the Anabaptists). He read to pupils here (among others, Walter Charlton, MD, was his pupil).

He has said oftentimes that the first rise, or hint of his rising, was from going accidentally a-coursing of a hare: where an ingenious gentleman of good quality falling into discourse with him, and finding him to have a very good wit, told him that he would never get any considerable preferment by continuing in the university; and that his best way was to betake himself to some lord's or great person's house that had good benefices to confer. Said Mr J. Wilkins, 'I am not known in the world; I know not to whom to address myself upon such a design.' The gentleman replied, 'I will recommend you myself,' and did so, to (as I think) Lord Viscount Say and Seae, where he stayed with very good liking till the late civil wars, and then he was chaplain to His Highness the Prince Elector Palatine of the Rhine, with whom he went (after the peace concluded in Germany) and was well preferred there by his highness. He stayed there not above a year.

After the visitation at Oxford by the Parliament, he got to be warden of Wadham College. In 1656 married to Robina the relict of Dr Peter French, canon of Christchurch, Oxford, and sister to Oliver, then Lord Protector, who made him in 1659 master of Trinity College in Cambridge (in which place he revived learning by strict examinations at elections: he was much honoured there, and heartily loved by all); where he continued till 1660 (the restoration of His Majesty). Then he was minister of St Laurence Jewry Church in London; and was after Dean of Ripon in Yorkshire. His friend, Seth Ward, DD, being made Bishop of Exeter, he was made there dean, and in 1668 by the favour of George, Duke of Buckingham, was made Bishop of Chester; and was extremely well beloved in his diocese. In 1672 he died of the stone. He left a legacy of £400 to the Royal Society, and had he been able would have given more. He was no great read man; but one of much and deep thinking, and of a working head; and a prudent man as well as ingenious. He was one of Seth, Lord Bishop of Salisbury's most intimate friends. He was a lusty, strong grown, well set, broad shouldered person, cheerful and hospitable.

He was the principal reviver of experimental philosophy at Oxford, where he had weekly an experimental philosophical club, which began 1649, and was the incunabula of the Royal Society. When he came to London, they met at the Bull-head tavern in Cheapside (e.g. 1658, 1659, and after) till it grew too big for a club, and so they came to Gresham College parlour.

THOMAS WILLIS
1621–75

Thomas Willis, M. born at Great Bedwyn in Wiltshire, 2nd January 1621. His father was steward to Sir Walter Smyth there, and had been sometime a scholar at St John's College in Oxford.

He studied chemistry in Peckwater Inn chamber. He was in those days very mathematical, and I have heard him say his genius lay more to mathematics than chemistry.

His father was steward to Sir John Smyth; and had a little estate at Ivy Hinksey, where my lady Smyth died.

He went to school to Mr (Edward) Sylvester in Oxford, over the meadows, where he aired his muse, and made good exercise. About 1657, riding towards Brackley to a patient, his way led him through Astrop, where he observed the stones in the little rill were discoloured of a kind of *Crocus Martis* colour; thought he, this may be an indication of iron; he gets galls, and puts some of the powder into the water, and immediately it turned blackish; then said he, 'I'll not send my patients now so far as Tunbridge,' and so he in a short time brought these waters into vogue, and has enriched a poor obscure village. He was middle stature: dark red hair (like a red pig): stammered much.

He was first servitor to Dr (Thomas) Iles, one of the canons of Christ Church whose wife was a knowing woman in physic and surgery, and did many cures. Tom Willis then wore a blue livery-cloak, and studied at the lower end of the hall, by the hall-door; was pretty handy, and his mistress would oftentimes have him assist her in making of medicines. This did him no hurt, and allured him on.

EDWARD WRIGHT
1558?–1615

Mr Edward Wright: he was of Caius College in Cambridge.

He published his book, quarto, entitled: 'Certain errors in navigation detected and corrected by Edward Wright, with many additions that were not in the former edition as appear in the next pages, London, 1610'.

It is dedicated to the high and mighty Henry, Prince of Wales, etc. In the epistle dedicatory he makes mention of a goodly and royal ship that His Highness lately built, and that since His Highness coming into England that the 'art of navigation has been much advanced here as well in searching the north-east and north-west passages as also in discovering the sea coasts and inlands of Virginia, Newfoundland, Greenland, and of the north new-land as far as Hakluyt's headland, within nine degrees of the pole, also of Guyana and diverse parts and islands of the East Indies, yea, and some parts also of the south continent discovered by Sir Richard Hawkins.' He read mathematics to Prince Henry; and Sir Jonas Moore had the wooden sphere in the Tower, which was contrived by Mr Wright for the more easy information of the prince.

Amongst Mr Laurence Rooke's papers (left with Seth Ward, Lord Bishop of Salisbury) I found *Hypothesis stellarum fixarum a Edm. Wright*,[75] three sheets, of his own handwriting, in folio. I deposited it in the Royal Society, but Mr R. Hooke says that it is printed in a book by itself, which see.

In his preface to the reader he says that 'the errors I have in the following treatise laboured to reform to the utmost (yea, rather beyond the utmost) of my poor ability, neglecting in the mean time other studies and courses that might have been more beneficial to me: which may argue my good will to have proceeded further to the amendment of such other faults and imperfections as yet remain besides those already specified.' It appears by his preface that his worth was attended by a great deal of envy.

He was in the voyage of the right honourable the Earl of Cumberland in the year 1589. He 'devised the seaman's rings for the present finding out both of the variation of the needle and time of the day at one instant without any further trouble of using any other instrument, and has further showed how by the sun's point of the compass (or magnetic azimuth) and altitude given by observation the variation may be found either mechanically or by the doctrine of triangles and arithmetical calculation.' John Collins says that he happened upon the logarithms and did not know it, as may be seen in his *Errors*: and Mr Robert Norwood says to the reader in his *Trigonometry* 'neither is Mr Edward Wright to be forgotten though his endeavours were soonest prevented,' speaking of the logarithms.

Mr Edward Wright was of Caius College, in Cambridge. He was one of the best mathematicians of his time; and the *then* new way of sailing, which yet goes by the name of 'sailing by Mr Mercator's chart' was purely his invention, as plainly does and may appear in his learned book called *Wright's Errors in Navigation*, in quarto printed... Mr Mercator brought this invention in fashion beyond seas.

NOTES

1. i.e. got into debt.
2. i.e. a boaster.
3. Gold coins.
4. 'Completely absorbed' (Latin).
5. 'Behind red hair is not a soul without bitterness' (Latin).
6. A type of entrance to an underground mine that is horizontal or nearly horizontal.
7. Leaves a hard deposit on anything left in water.
8. A tap.
9. 'Base-born', i.e. illegitimate.
10. i.e. with doors barred.
11. i.e. interested in chemistry.
12. Legal term: a half.
13. Testicles.
14. 'On the spot' (French).
15. Orally, literally 'with living voice' (Latin).
16. Exposed.
17. Ride on a well-equipped horse.
18. 'They would smite him undismayed' (Latin).
19. Aubrey's lives were written for Sir Anthony Wood and here he addresses him directly.
20. 'We cannot all do everything' (Latin).
21. 'Let the astronomers work it out' (Latin).
22. Clarke notes here, 'Perhaps because the letters ended in tridents.'
23. 'The Use of the Globes' (Latin).
24. 'Nothing is made out of nothing' (Latin).
25. 'Nothing' (Latin).
26. Skin cancer, literally 'touch me not' (Latin). The Latin term was given because the cancerous growths were thought to be incurable and therefore not to be treated.
27. The Great Fire of London, 1666.
28. 'Key to Mathematics' (Latin).
29. Modern authors.
30. 'The Circulation of the Blood' (Latin).
31. 'On the Reproduction of Animals' (Latin).
32. Prescription.
33. A long cloth covering most of the horse's back.
34. 'Spinal column' (Latin).
35. i.e. people who have married for wealth.
36. 'They are completely on the wrong track' (Latin).

37. Cuckolds, men whose wives have been unfaithful, were often described as having horns on their head.
38. i.e. northern.
39. Nitric acid.
40. Red ochre.
41. 'In the end' (Latin).
42. Support.
43. i.e. the title page.
44. 'What he sought' (Latin).
45. i.e. a beaver hat.
46. 'In the time of Elizabeth' (Latin).
47. Probably Thomas Henshawe.
48. i.e. a way of working out longitude while at sea.
49. 'But I scarcely believe it' (Latin).
50. 'Not without God's help' (Latin).
51. 'Legate' (Latin).
52. 'With the attached chapel of Bartlesdon in the same county' (Latin).
53. i.e. his horoscope.
54. i.e. his horoscope.
55. Patron.
56. Widow.
57. 'Take nobody's word for it' – the Latin motto of the Royal Society.
58. 'Sluggish' (Latin).
59. 'The mystery of the Beast was discovered' (Latin).
60. 'Whether the Pope was Anti-Christ' (Latin).
61. 'On the Nature of Animals' (Latin).
62. A centre bar.
63. Soreness or bleariness of the eyes.
64. Blemish.
65. 'In the time of Charles I and II' (Latin).
66. Shabbo is a skin disease of sheep, cured by the application of tar.
67. Contemporary.
68. i.e. scandalously accused of.
69. An agnostic.
70. Job 30: 31. 'My harp also is turned to mourning, and my organ into the voice of them that weep' (Latin).
71. 'Some kind of poison' (Latin).
72. Riddles.
73. 'With one voice' (Latin).
74. i.e. a bishop's vestments.
75. 'A hypothesis of fixed stars, by Edward Wright' (Latin).

BIOGRAPHICAL NOTE

Born near Malmesbury in Wiltshire in 1626, John Aubrey was the oldest surviving son of a well-off gentry family. He was educated at Malmesbury Grammar School under Robert Latimer, and it was here he made the acquaintance of Thomas Hobbes, about whom he would later write.

He went on to enter Trinity College, Oxford, but his education was interrupted by the English Civil War. In 1646 he became a student of the Middle Temple.

Aubrey was an antiquary and in 1648, was the first person to discover the ruins of Avebury. He devoted much time to archaeological research and became one of the original fellows of the Royal Society in 1662. Throughout his life, Aubrey was acquainted with the most celebrated writers, scientists and politicians of his day, along with a large number of other distinguished figures *Miscellanies* (1696) was the only work he published during his lifetime. His *Natural History of Wiltshire* was made available in 1847, but he is best remembered for his *Lives of Eminent Men* (sometimes referred to as *Brief Lives*), a series of biographical notes, court gossip and scurrilous anecdotes collected over a number of years.

Aubrey died of apoplexy in June 1697 while travelling, and was buried in the churchyard at St Mary Magdalene, Oxford.

HESPERUS PRESS

Hesperus Press is committed to bringing near what is far – far both in space and time. Works written by the greatest authors, and unjustly neglected or simply little known in the English-speaking world, are made accessible through new translations and a completely fresh editorial approach. Through these classic works, the reader is introduced to the greatest writers from all times and all cultures.

For more information on Hesperus Press, please visit our website: **www.hesperuspress.com**

SELECTED TITLES FROM HESPERUS PRESS

Author	*Title*	*Foreword writer*
Pietro Aretino	*The School of Whoredom*	Paul Bailey
Pietro Aretino	*The Secret Life of Nuns*	
Jane Austen	*Lesley Castle*	Zoë Heller
Jane Austen	*Love and Friendship*	Fay Weldon
Honoré de Balzac	*Colonel Chabert* •	A.N. Wilson
Charles Baudelaire	*On Wine and Hashish*	Margaret Drabble
Giovanni Boccaccio	*Life of Dante*	A.N. Wilson
Charlotte Brontë	*The Spell*	
Emily Brontë	*Poems of Solitude*	Helen Dunmore
Mikhail Bulgakov	*Fatal Eggs*	Doris Lessing
Mikhail Bulgakov	*The Heart of a Dog*	A.S. Byatt
Giacomo Casanova	*The Duel*	Tim Parks
Miguel de Cervantes	*The Dialogue of the Dogs*	Ben Okri
Geoffrey Chaucer	*The Parliament of Birds*	
Anton Chekhov	*The Story of a Nobody*	Louis de Bernières
Anton Chekhov	*Three Years*	William Fiennes
Wilkie Collins	*The Frozen Deep*	
Joseph Conrad	*Heart of Darkness* •	A.N. Wilson
Joseph Conrad	*The Return*	Colm Tóibín
Gabriele D'Annunzio	*The Book of the Virgins*	Tim Parks
Dante Alighieri	*The Divine Comedy: Inferno* •	
Dante Alighieri	*New Life*	Louis de Bernières
Daniel Defoe	*The King of Pirates*	Peter Ackroyd
Marquis de Sade	*Incest*	Janet Street-Porter
Charles Dickens	*The Haunted House*	Peter Ackroyd
Charles Dickens	*A House to Let*	
Fyodor Dostoevsky	*The Double*	Jeremy Dyson
Fyodor Dostoevsky	*Poor People*	Charlotte Hobson
Alexandre Dumas	*One Thousand and One Ghosts*	

George Eliot	*Amos Barton*	Matthew Sweet
Henry Fielding	*Jonathan Wild the Great*	Peter Ackroyd
F. Scott Fitzgerald	*The Popular Girl*	Helen Dunmore
Gustave Flaubert	*Memoirs of a Madman*	Germaine Greer
Ugo Foscolo	*Last Letters of Jacopo Ortis*	Valerio Massimo Manfredi
Elizabeth Gaskell	*Lois the Witch*	Jenny Uglow
Théophile Gautier	*The Jinx*	Gilbert Adair
André Gide	*Theseus*	
Johann Wolfgang von Goethe	*The Man of Fifty*	A.S. Byatt
Nikolai Gogol	*The Squabble*	Patrick McCabe
E.T.A. Hoffmann	*Mademoiselle de Scudéri*	Gilbert Adair
Victor Hugo	*The Last Day of a Condemned Man*	Libby Purves
Joris-Karl Huysmans	*With the Flow*	Simon Callow
Henry James	*In the Cage*	Libby Purves
Franz Kafka	*Metamorphosis*	Martin Jarvis
Franz Kafka	*The Trial*	Zadie Smith
John Keats	*Fugitive Poems*	Andrew Motion
Heinrich von Kleist	*The Marquise of O–*	Andrew Miller
Mikhail Lermontov	*A Hero of Our Time*	Doris Lessing
Nikolai Leskov	*Lady Macbeth of Mtsensk*	Gilbert Adair
Carlo Levi	*Words are Stones*	Anita Desai
Xavier de Maistre	*A Journey Around my Room*	Alain de Botton
André Malraux	*The Way of the Kings*	Rachel Seiffert
Katherine Mansfield	*Prelude*	William Boyd
Edgar Lee Masters	*Spoon River Anthology*	Shena Mackay
Guy de Maupassant	*Butterball*	Germaine Greer
Prosper Mérimée	*Carmen*	Philip Pullman
Sir Thomas More	*The History of King Richard III*	Sister Wendy Beckett

Sándor Petőfi	*John the Valiant*	George Szirtes
Francis Petrarch	*My Secret Book*	Germaine Greer
Luigi Pirandello	*Loveless Love*	
Edgar Allan Poe	*Eureka*	Sir Patrick Moore
Alexander Pope	*The Rape of the Lock* and *A Key to the Lock*	Peter Ackroyd
Antoine-François Prévost	*Manon Lescaut*	Germaine Greer
Marcel Proust	*Pleasures and Days*	A.N. Wilson
Alexander Pushkin	*Dubrovsky*	Patrick Neate
Alexander Pushkin	*Ruslan and Lyudmila*	Colm Tóibín
François Rabelais	*Pantagruel*	Paul Bailey
François Rabelais	*Gargantua*	Paul Bailey
Christina Rossetti	*Commonplace*	Andrew Motion
George Sand	*The Devil's Pool*	Victoria Glendinning
Jean-Paul Sartre	*The Wall*	Justin Cartwright
Friedrich von Schiller	*The Ghost-seer*	Martin Jarvis
Mary Shelley	*Transformation*	
Percy Bysshe Shelley	*Zastrozzi*	Germaine Greer
Stendhal	*Memoirs of an Egotist*	Doris Lessing
Robert Louis Stevenson	*Dr Jekyll and Mr Hyde*	Helen Dunmore
Theodor Storm	*The Lake of the Bees*	Alan Sillitoe
Leo Tolstoy	*The Death of Ivan Ilych*	
Leo Tolstoy	*Hadji Murat*	Colm Tóibín
Ivan Turgenev	*Faust*	Simon Callow
Mark Twain	*The Diary of Adam and Eve*	John Updike
Mark Twain	*Tom Sawyer, Detective*	
Oscar Wilde	*The Portrait of Mr W.H.*	Peter Ackroyd
Virginia Woolf	*Carlyle's House and Other Sketches*	Doris Lessing
Virginia Woolf	*Monday or Tuesday*	Scarlett Thomas
Emile Zola	*For a Night of Love*	A.N. Wilson